Test Bank to Accompany

HUMAN ANATOMY & PHYSIOLOGY

SECOND EDITION

SOLOMON
SCHMIDT
ADRAGNA

Richard R. Schmidt, Ph.D.

Saunders College Publishing
Philadelphia, Fort Worth, Chicago, San Francisco
Montreal, Toronto, London, Sydney, Tokyo

Copyright ©1990 by Saunders College Publishing, a division of Holt, Rinehart and Winston.

All rights reserved. No part of this publication may be reproduced or transmitted in any form or by any means, electronic or mechanical, including photocopy, recording or any information storage and retrieval system, without permission in writing from the publisher.

Requests for permission to make copies of any part of the work should be mailed to: Permissions, Holt, Rinehart and Winston, Inc., Orlando, Florida 32887.

Printed in the United States of America.

Solomon/Schmidt/Adragna: Test Bank to accompany HUMAN ANATOMY AND PHYSIOLOGY, 2nd edition

ISBN # 0-03-033117-X

012 046 987654321

TEST BANK for CHAPTER 1: INTRODUCING THE HUMAN BODY

INSTRUCTIONS: COMPLETE THE FOLLOWING

1. The medical specialty concerned with tumors and cancer is called:

ANSWER: oncology

2. The medical specialty concerned with the eyes and eye disorders is:

ANSWER: ophthamology

3. The medical specialty concerned with foot aliments is called:

ANSWER: podiatry

4. The medical specialty concerned with pregnancy and child birth is called:

ANSWER: obstetrics

5. Catabolism and anabolism are the two phases of:

ANSWER: metabolism

6. The automatic tendency to maintain a relatively constant internal environment is called:

ANSWER: homeostasis or homeodynamics

7. The basic units of all matter are the:

ANSWER: atoms

8. Each cell consists of specialized parts called:

ANSWER: organelles

9. The organelle that surrounds the cell and regulates substances passing into and out of the cell is called the:

ANSWER: plasma membrane

10. The organelle that serves as the information and control center of the cell is the:

ANSWER: nucleus

11. A group of closely associated cells specialized to perform a particular function is called a:

ANSWER: tissue

12. Various tissues may be organized into structures called:

ANSWER: organs

13. A coordinated group of tissues and organs makes up a:

ANSWER: body system

14. The body systems make up the complex living:

ANSWER: organism

15. The three types of muscle are skeletal muscle, smooth muscle, and _____ muscle.

ANSWER: cardiac

16. Skin, nails and hair make up part of the _____ system.

ANSWER: integumentary

17. The endocrine glands secrete substances called:

ANSWER: hormones

18. The _____ system is the transportation system of the body.

ANSWER: circulatory

19. The heart, blood, and blood vessels constitute the _____ system.

ANSWER: cardiovascular

20. The cardiovascular system and the lymphatic system are included in the _____ system

ANSWER: circulatory

21. The air passageways and the lungs are included in the _____ system.

ANSWER: respiratory

22. Metabolic waste disposal in mainly the job of the _____ system.

ANSWER: urinary

23. The kidneys and ureters are include in the _____ system.

ANSWER: urinary

24. The ovaries, uterus and penis are included in the _____ system.

ANSWER: reproductive

25. Feedback systems of the body are often referred to as _____ systems.

ANSWER: biofeedback

26. Most homeostatic mechanisms in the human body are _____ feedback systems.

ANSWER: negative

27. Many _____ feedback systems are vicious cycles that lead to disruption of steady states and even to death.

ANSWER: positive

28. Human body temperature is maintained by _____ feedback mechanisms.

ANSWER: negative

29. When the glucose level in the blood begins to fall, the pancreas releases a hormone called:

ANSWER: glucagon

30. The human body exhibits _____ symmetry.

ANSWER: bilateral

31. The term _____ means to be located on the same side of the body.

ANSWER: ipsilateral

32. The term _____ means to be located on the opposite side of the body.

ANSWER: contralateral

33. The body may be divided into an _____ portion and an appendicular portion.

ANSWER: axial

34. A _____ plane divides the body into superior and inferior parts.

ANSWER: transverse

35. A _____ plane divides the body into right and left halves.

ANSWER: sagittal

36. The body cavities contain the internal organs, or _____.

ANSWER: viscera

37. The thoracic and abdominopelvic cavities are separated by a broad muscle called the_____.

ANSWER: diaphragm

INSTRUCTIONS: TRUE / FALSE

38. (T/F): Nuclear scanning and CT scanning are techniques utilized to visualize body parts clinically.

ANSWER: T

39. (T/F): A frontal plane is also known as a coronal plane.

ANSWER: T

40. (T/F): A frontal plane lies at right angles to a sagittal plane.

ANSWER: T

41. (T/F): The right leg and right arm are ipsilateral structures.

ANSWER: T

43. (T/F): The right leg and left leg are contralateral structures.

ANSWER: T

44. (T/F): The wrist is distal to the fingers.

ANSWER: F

45. (T/F): The ankle is proximal to the toes.

ANSWER: T

46. (T/F): The vertebral column is posterior to the liver.

ANSWER: T

47. (T/F): The head, neck, and trunk make up the appendicular portion of the body.

ANSWER: F

48. (T/F): The thorax, abdomen and pelvis constitute the torso.

ANSWER: T

49. (T/F): The limbs constitute the appendicular portion
of the body.

ANSWER: T

50. (T/F): The term occipital refers to the posterior aspect
of the head.

ANSWER: T

51. (T/F): The term tarsal refers to ankle.

ANSWER: T

52. (T/F): The term inguinal refers to the groin.

ANSWER: T

53. (T/F): The two principal body cavities are the dorsal cavity
and the ventral cavity.

ANSWER: T

54. (T/F): The ventral cavity is divided into the thoracic and
abdominopelvic cavities.

ANSWER: T

INSTRUCTIONS: ANSWER EACH QUESTION ACCORDING TO THE FOLLOWING KEY:

(A) Only 1 is correct
(B) Only 2 is correct
(C) Both are correct
(D) Neither are correct

55. Characteristic of living things:

1. movement
2. responsiveness

ANSWER: C

TB.1:7

56. *Characteristic of living things:

    1. growth and development
    2. metabolism

ANSWER: C

57. Characteristic of living things:

    1. adaptation
    2. reproduction

ANSWER: C

58. One of the main types of tissue in the body:

    1. connective tissue
    2. epithelial tissue

ANSWER: C

59. One of the main types of tissue in the body:

    1. nervous tissue
    2. cardiac muscle

ANSWER: A

60. Ventral cavity:

    1. cranial cavity
    2. thoracic cavity

ANSWER: A

61. Dorsal cavity:

    1. cranial cavity
    2. abdominopelvic cavity

ANSWER: A

62. A principal body cavity:

    1. dorsal cavity
    2. ventral cavity

ANSWER: C

63. Located in abdominal cavity:

    1. spleen
    2. kidneys

ANSWER: C

64. Located in thoracic cavity:

    1. lungs
    2. heart

ANSWER: C

65. Division of thoracic cavity:

    1. heart
    2. mediastinum

ANSWER: B

66. Located in pelvic cavity:

    1. part of large intestine
    2. uterus

ANSWER: C

67. Techinque used clinically to visualize body parts:

    1. MRI
    2. thermography

ANSWER: C

68. Which of the following are correct?

    1. fingers are distal to the wrist
    2. knee is proximal to the ankle

ANSWER: C

69. Which of the following are correct?

    1. knee is proximal to the hip
    2. wrist is proximal to the elbow

ANSWER: D

70. Which of the following is correct?

    1. transverse plane is perpendicular to a sagittal plane
    2. coronal plane is perpendicular to a sagittal plane

ANSWER: C

71. Portion of the torso:

    1. head
    2. upper limb

ANSWER: D

72. Portion of the torso:

    1. abdomen
    2. thorax

ANSWER: C

73. Portion of the torso:

    1. thorax
    2. pelvis

ANSWER: C

74. Anatomical position:

    1. palms facing forward
    2. standing erect with arms at side of body

ANSWER: C

75. When homeostatic mechanisms are unable to restore the steady state, stress may lead to:

    1. death
    2. malfunction

ANSWER: C

TEST BANK for CHAPTER 2: THE CHEMISTRY OF LIFE

INSTRUCTIONS: TRUE - FALSE

1. (T/F): The smallest unit of an element that retains the characteristic chemical properties of that element is called an atom.

ANSWER: T

2. (T/F): An element may be composed of one or more different atoms.

ANSWER: T

3. (T/F): Carbon is the most plentiful element in the human body.

ANSWER: F

4. (T/F): Oxygen is the most plentiful element in the human body.

ANSWER: T

5. (T/F): Collectively, electrons and protons are called nucleons.

ANSWER: F

6. (T/F): Collectively, protons and neutrons are called necleons.

ANSWER: T

7. (T/F): Electrons and protons are located within the nucleus of an atom.

ANSWER: F

8. (T/F): The nucleus of an atom contains protons and neutrons.

ANSWER: T

9. (T/F): Electrons determine the atomic number of an atom.

ANSWER: F

10. (T/F): Together, protons and neutrons determine the atomic number of an atom.

ANSWER: F

11. (T/F): Protons determine the atomic number of an atom.

ANSWER: T

12. (T/F): Each atom has a characteristic number of protons in its nucleus.

ANSWER: T

13. (T/F): The atomic number of hydrogen is 1.

ANSWER: T

14. (T/F): An atom with 6 protons in its nucleus also has an atomic number of 6.

ANSWER: T

15. (T/F): An atom with six protons and six neutrons in its nucleus has an atomic weight of 12.

ANSWER: T

16. (T/F): Atoms of an element with different numbers of protons are called isotopes.

ANSWER: F

17. (T/F): Atoms of an element with different numbers of neutrons are called isotopes.

ANSWER: T

18. (T/F): Carbon-14 has eight neutrons in its nucleus.

ANSWER: T

19. (T/F): Some isotopes are stable while others break down and release high-energy radiation.

ANSWER: T

20. (T/F): Carbon-14 is a stable isotope.

ANSWER: F

21. (T/F): Carbon-14 is a radioisotope or radioneuclide.

ANSWER: T

22. (T/F): Electrons are positively charged particles that whirl around the atomic nucleus.

ANSWER: F

23. (T/F): An isolated atom has the same number of protons and electrons and is said to be electrically neutral.

ANSWER: T

24. (T/F): Two or more atoms combine chemically to form a molecule.

ANSWER: T

25. (T/F): A molecule is composed of at least two or more chemically combined atoms.

ANSWER: F

26. (T/F): A molecule is composed of at least two different types of chemically combined atoms.

ANSWER: T

27. (T/F): A structural formula for a molecule shows the arrangement of the atoms in that molecule.

ANSWER: T

28. (T/F): The elctrons in the outer shell of an atom are called its valence electrons.

ANSWER: T

29. (T/F): Gases that do not normally participate in chemical reactions are called inert gases.

ANSWER: T

30. (T/F): Nonmetals tend to behave as electron acceptors.

ANSWER: T

INSTRUCTIONS: ANSWER EACH QUESTION ACCORDING TO THE FOLLOWING KEY:

    (A) Only 1 is correct
    (B) Only 2 is correct
    (C) Both are correct
    (D) Neither are correct

31. An atom becomes electrically charged when it:

    1. gains electrons
    2. loses electrons

ANSWER: C

32. In covalent bonds, there is no:

    1. gain of electrons
    2. loss of electrons

ANSWER: C

33. Required for the breakdown of sucrose:

    1. water
    2. enzyme

ANSWER: C

34. Water:

    1. accounts for about 80% of the body weight of an adult
    2. forms about 90% of blood plasma

ANSWER: B

35. Water:

    1. helps maintain the constant temperature of the body
    2. is considered the most important biological solvent

ANSWER: C

36. Regarding water:

    1. its polarity allows it to be a strong solvent
    2. its electrical charges exert a force upon compounds immersed in it

ANSWER: C

37. An acid in solution:

    1. dissociates
    2. produces hydrogen ions

ANSWER: C

38. Acids:

    1. are often referred to as proton donors
    2. dissociate in solution

ANSWER: C

39. Considered a weak acid:

    1. hydrochloric acid
    2. carbonic acid

ANSWER: B

40. A base is:

    1. also called an alkali
    2. dissociates in solution

ANSWER: C

41. Measurement of pH:

    1. is on a scale of 0 to 14
    2. is a method for determining the relative acidity or alkalinity of a solution

ANSWER: C

42. The more acidic a solution is:

    1. the more hyderogen ions are present
    2. the higher the pH will be

ANSWER: A

43. A buffer consists of:

    1. weak acid
    2. a salt of the weak acid present

ANSWER: C

44. Organic compounds:

    1. contain hydrogen
    2. are complec compounds containing carbon

ANSWER: C

45. Organic compounds may contain:

    1. nitrogen
    2. phosphorus

ANSWER: C

46. Carbon:

    1. has four electrons in its outer shell
    2. forms very strong covalent bonds

ANSWER: C

47. Carbon:

    1. may form a double bond
    2. is not found in organic compounds

ANSWER: A

TB.2:7

48. Carbon atoms:

    1. may enter into the formation of polymers
    2. does not form double bonds

ANSWER: A

49. Importants classes of organic compounds include:

    1. carbohydrates
    2. nucleic acids

ANSWER: C

50. Inculded among the classes of important organic compounds are:

    1. proteins
    2. lipids

ANSWER: C

INSTRUCTIONS: COMPLETE THE FOLLOWING:

51. Negatively charged ions are called:

ANSWER: anions

52. Positively charged ions are called:

ANSWER: cations

53. Bonds between ions are called:

ANSWER: ionic bonds

54. Starches, sugars, and cellulose are exampoles of:

ANSWER: carbohydrates

55. Carbohydrates may be classified as monosaccharides, disaccarhides, or:

ANSWER: polysaccharides

56. Monosaccharides are also known as simple:

ANSWER: sugars

57. Monosaccharides contain from three to seven:

ANSWER: carbon atoms

58. The most common monosaccharide in living organisms is:

ANSWER: glucose

59. Two monosaccharides may combine to form a:

ANSWER: disaccharide

60. In a disaccharide, two monosaccharides are joined by a:

ANSWER: glycosidic linkage

61. A very weak chemical bond between an already bonded hydrogen atom and a negatively charged atom such as oxygen or nitrogen is called a:

ANSWER: hydrogen bond

62. The polysaccharide responsible for the rigidity of the plant cell wall is called:

ANSWER: cellulose

63. The class of compounds that yield the most fuel energy by weight are known as:

ANSWER: neutral lipids

64. A three carbon alcohol that contains three hydroxyl groups is called:

ANSWER: glycerol

65. A type of fatty acid that contains the maximum number of hydrogen atoms chemically possible is known as a:

ANSWER: saturated fatty acid

66. Three fatty acids attached to a glycerol is a:

ANSWER: triglyceride

67. A triglyceride containing three unsaturated fatty acids is referred to as a:

ANSWER: polyunsaturated fat

68. Glycerol that is chemically combined with two fatty acids and a phosphate group is referred to as:

ANSWER: phosphatidic acid

69. The fatty acid end of a phospholipid molecule is repelled by water and is thus said to be:

ANSWER: hydrophobic

70. The base end of a phospholipid molecule is attracted to water and is thus said to be:

ANSWER: hydrophilic

71. The most abundant steroid in the human body is called:

ANSWER: cholesterol

72. Three types of RNA include, messenger RNA, transfer RNA, and:

ANSWER: ribosomal RNA

73. The compound which contains all of the information necessary for making all of the proteins required by the organism is:

ANSWER: deoxyribonucleic acid (DNA)

74. The nitrogenous bases of DNA may be classified as purines or:

ANSWER: pyrimidines

75. The purine bases of DNA are adenine and:

ANSWER: guanine

76. The pyrimidine bases of DNA are cytosine and:

ANSWER: thymine

77. The pyrimidine bases of RNA are cytosine and:

ANSWER: uracil

78. The covalent bond linking two amino acids is called a:

ANSWER: peptide bond

79. Two amino acids combine chemically to form a:

ANSWER: dipeptide

80. The three dimensional shape of a polypeptide is refferred to as its:

ANSWER: conformation

81. The sequence of amino acids in a peptide chain is referred to as the:

ANSWER: primary structure

82. Four levels of organization can be distinguished in a protein molecule - primary, secondary, tertiary, and:

ANSWER: quaternary structure

83. Organic catylists that greatly increase the speed of chemical reactions without being consumed themselves are called:

ANSWER: enzymes

84. A smaller, nonprotein molecule required for certain enzymes to function is called a:

ANSWER: cofactor

85. An organic, nonpolypeptide compound serving as a cofactor is called a:

ANSWER: coenzyme

86. Enzyme inhibition may be either reversible or:

ANSWER: irreversible

TB.3:1

TEST BANK for CHAPTER 3: THE CELL: BUILDING BLOCK OF THE BODY

INSTRUCTIONS: SELECT THE MOST APPROPRIATE ANSWER FROM COLUMN B. THE SAME ANSWER MAY BE USED ONCE, MORE THAN ONCE, OR NOT AT ALL.

Column A	Column B
1. ( ) microtubules	A. resposible for movement of materials on surface of some tissues
2. ( ) ribosomes	B. membranous sacs containing enzymes that break down ingested materials, secretions, and wastes
3. ( ) Golgi complex	C. modifies and packages secreted proteins
4. ( ) nucleolus	D. contains nine microtubular triplets
5. ( ) centriole	E. located within nucleus and contains RNA and protein
6. ( ) cilia	F. some are free and some are attached to the ER
7. ( ) lysosome	G. hollow tubes composed of the protein, tubulin

ANSWERS: 1. (G), 2. (F), 3. (C), 4. (E), 5. (D), 6. (A), 7. (B)

TB.3:2

INSTRUCTIONS: ANSWER EACH QUESTION ACCORDING TO THE FOLLOWING KEY:

    (A) Only 1 is correct
    (B) Only 2 is correct
    (C) Both are correct
    (D) Neither are correct

8. Regarding mitochondria:

    1. tiny poerplants of the cell
    2. a duble membrane sac

ANSWER: C

9. Mitochondrial cristae:

    1. folds of the outer mitochondrial membrane
    2. project into the central compartment, or matrix

ANSWER: B

10. Mitochondrial cristae:

    1. contain some enzymes for cellular respiration
    2. increase surface area

ANSWER: C

11. Regarding the peroxisomes:

    1. membrane bound organelles
    2. contain the enzyme, catalase

ANSWER: C

12. Intermediate filaments:

    1. composed of polypeptides
    2. structure is uniform from cell to cell

ANSWER: A

13. Centriole:

    1. each cell possesses two such organelles
    2. major function is to package proteins for secretion

ANSWER: A

14. Microtubules:

    1. composed mainly of either actin or myosisn
    2. function primarily in cell movement

ANSWER: D

15. Have a 9 + 2 arrangement of microtubules:

    1. cilia
    2. flagella

ANSWER: C

16. Its base consists of a basal body:

    1. cilium
    2. flagellum

ANSWER: C

17. Regarding the basal body:

    1. forms the base of a cilium
    2. constructed similar to a centriole

ANSWER: C

18. Nucleus:

    1. may contain one or more nucleoli
    2. enveloped by an impermeable nuclear membrane

ANSWER: A

19. The plasma membrane:

    1. selectively permeable membrane
    2. a fluid lipid bilayer

ANSWER: C

20. Smooth endoplasmic reticulum:

    1. main site of fatty acid metabolism
    2. principal site of protein synthesis

ANSWER: A

21. Smooth endoplasmic reticulum:

    1. produces steroids in some cells
    2. stores calcium in muscle cells

ANSWER: C

22. Regarding the smooth ER of liver cells:

    1. important in detoxification
    2. increases when exposed to phenobarbital

ANSWER: C

23. Regarding the Golgi complex:

    1. consists largely of a network of microfliaments
    2. functions primarily in lipid metabolism

ANSWER: D

24. Regarding the endoplasmic reticulum:

    1. has cavities called cisternae
    2. may have ribosomes attached to it

ANSWER: C

25. Regarding lysosomes:

    1. contain digestive enzymes
    2. are involved in the "self-destruct" mechanism of dying cells

ANSWER: C

26. Regarding the Golgi complex:

    1. may renew its membranes evry 30 minutes in actively secreting cells
    2. may package its products in secretory vesicles

ANSWER: C

27. Regarding diffusion:

   1. net movement of particles down a consentration gradient
   2. rate is not influenced by particle size or shape

ANSWER: A

28. Regarding the process of diffusion:

   1. its rate is not influenced by temperature
   2. its rate is influenced by electrical charge of the particles involved

ANSWER: B

29. Regarding the process of facilitated diffusion:

   1. dependent on energy input by the cell
   2. involves a carrier protein

ANSWER: B

30. Transported across membranes via facilitated diffusion:

   1. amino acids
   2. glucose

ANSWER: C

31. Permease:

   1. an integral protein
   2. a carrier protein

ANSWER: C

32. Active transport:

   1. requires cellular energy
   2. moves materials against a concentration gradient

ANSWER: C

33. The sodium-potassium pump:

    1. an example of active transport
    2. pumps potassium ions out of the cell

ANSWER: A

34. A form of endocytosis:

    1. phagocytosis
    2. pinocytosis

ANSWER: C

35. Regarding the process of pinocytosis:

    1. referred to as "cell drinking"
    2. a type of exocytosis

ANSWER: A

36. Receptor-mediated endocytosis:

    1. involves ligand binding
    2. receptor proteins returned to the membrane for reuse

ANSWER: C

INSTRUCTIONS: TRUE - FALSE

37. (T/F): Exocytosis and endocytosis require the expenditure of cellular energy.

ANSWER: T

38. (T/F): The length of the cell cycle is uniform from one cell type two another.

ANSWER: F

39. (T/F): Cell division involves mitosis and cytokinesis.

ANSWER: T

40. (T/F): Cytokinesis means division of the cytoplasm.

ANSWER: T

41. (T/F): The cell cycle is typically divided into five distinct phases.

ANSWER: F

42. (T/F): Cytokinesis generally takes place during prophase of the cell cycle.

ANSWER: F

43. (T/F): A cleavage furrow is typically observed as cytokinesis proceeds.

ANSWER: T

44. (T/F): Chalones are substances that stimulate cell division.

ANSWER: F

45. (T/F): Some growth factors may be found circulating in the blood.

ANSWER: T

46. (T/F): Chalones, in general, are non-specific in their action.

ANSWER: F

47. (T/F): The G_1 phase of the cell cycle occurs during interphase.

ANSWER: T

48. (T/F): Spindle fibers attach to the centromeres toward the end of prophase.

ANSWER: T

49. (T/F): The mitotic spindle forms during prophase.

ANSWER: T

50. (T/F): The chromatids separate during prophase.

ANSWER: F

INSTRUCTIONS: COMPLETE THE FOLLOWING:

51. The part of the cell cycle in which a cell spends most of its life is called:

ANSWER: interphase

52. The S phase is a subphase of the _____ portion of the cell cycle.

ANSWER: interphase

53. A term that literally means, 'cell eating', is:

ANSWER: phagocytosis

54. A cell ejects waste products or secretions by the process known as:

ANSWER: exocytosis

55. Included among the three types of endocytosis are phagocytosis, pinocytosis, and:

ANSWER: receptor-mediated endocytosis

56. Hydrostatic pressure may be utilized to push substances through cell membranes, a process termed:

ANSWER: filtration

57. A special type of diffusion in which water molecules move across a selectively permeable membrane in a direction where they are less concentrated is called:

ANSWER: osmosis

58. Glucose is transported across cell membranes by a process called:

ANSWER: facilitated diffusion

59. Amino acids are transported across cell membranes by a process called:

ANSWER: facilitated diffusion

60. The process of diffusion involves the net movement of particles down a:

ANSWER: concentration gradient

61. The energy source for diffusion is random molecular:

ANSWER: motion

62. Facilitated diffusion requires the use of a:

ANSWER: carrier protein

63. Microtubular pairs in cilia are connected to one another by proteins called:

ANSWER: dyneins

64. The major protein of microtubules is called:

ANSWER: tubulin

65. Actin and myosin are the principal proteins associated with tiny rodlike structures called:

ANSWER: microfilaments

66. The dense area of cytoplasm where centrioles are typically located is called the:

ANSWER: centrosome

67. Each centriole is composed of nine sets of three:

ANSWER: microtubules

68. Structures proposed to serve as conducting tracks along which substances move through the cytoplasm of the cell are called:

ANSWER: microtubules

69. The inner compartment of a mitochondrion is called its:

ANSWER: matrix

70. The folds of the inner mitochondrial membrane are called the:

ANSWER: cristae

71. Rheumatoid arthritis is thought to result in part from damage to cartilage cells in joints by enzymes that have been released from organelles called:

ANSWER: lysosomes

72. Cellular factories whose principal work is to assemble proteins are called:

ANSWER: ribosomes

73. Cavities formed by sheets of endoplasmic reticulum are called:

ANSWER: cisternae

74. Smooth endoplasmic reticulum lacks the presence of:

ANSWER: ribosomes

75. The site of ribosomal RNA synthesis is the:

ANSWER: nucleolus

76. The cellular structure which serves to propel sperm through the environment is the:

ANSWER: flagellum

77. Drug detoxification is the principal function of:

ANSWER: smooth endoplasmic reticulum

78. The tiny structures which perform specialized functions within a cell are collectively referred to as its:

ANSWER: organelles

79. The cell membrane is also called the:

ANSWER: plasma membrane

80. Proteins that penetrate into the hydrophobic regions of the lipid bilayer are designated as:

ANSWER: integral proteins

TEST BANK for CHAPTER 4: TISSUES: THE FABRIC OF THE BODY

INSTRUCTIONS: COMPLETE THE FOLLOWING:

1. The study of tissues is called:

ANSWER: histology

2. A type of tissues specailized for moving the body and its parts:

ANSWER: muscle tissue

3. A type of tissue that receives and transmits messages such that various parts of the body can communicate with one another:

ANSWER: nervous tissue

4. A cell junction characterized by disc-shaped points of contact is the:

ANSWER: spot desmosome

5. A type of cell junction characterized by a band of cell-to-cell adhesion extending around each cell:

ANSWER: belt desmosome

6. A thickened region of the plasma membranes of two adjacent cells forms a cell junction known as a:

ANSWER: desmosome

7. The two components of the basement membrane are the basal lamina and the:

ANSWER: reticular lamina

8. A type of epithelium in which the cells fasely appear to be layered is:

ANSWER: pseudostratified epithelium

9. Epithelium may be simple, pseudostratified, or:

ANSWER: startified

10. Epithelium may be classified as simple, stratified, or:

ANSWER: pseudostratified

11. Epithelium may be classified as stratified, pseudostratified, or:

ANSWER: simple

12. The lining of the body cavities consists of simple squamous epithelium called:

ANSWER: mesothelium

13. The epithelium lining the inner walls of blood and lymphatic vessels is called:

ANSWER: endothelium

14. As cells of the epidermis die they produce a tough, fibrous, water-proofing protein called:

ANSWER: keratin

15. A type of epithelium similar to stratified squamous, but also having the ability to stretch is:

ANSWER: transitional epithelium

16. Two main types of glands are endocrine and:

ANSWER: exocrine

17. A type of gland that lacks ducts:

ANSWER: endocrine

18. Endocrine glands produce substances called:

ANSWER: hormones

19. Exocrine glands may be unicellular or:

ANSWER: multicellular

20. A multicellular gland may be simple or:

ANSWER: compound

21. Alveolar glands are also known as:

ANSWER: acinar glands

22. Exocrine glands may be classified as mucous, serous, or:

ANSWER: mixed

23. Glands that release their secretions without loss of cytoplasm or damage to the cell are called:

ANSWER: merocrine glands

24. A type of gland in which the cell dies and becomes part of that secretion is referred to as a:

ANSWER: holocrine gland

25. Due to their mode of secretion, sebaceous glands of the skin are classified as:

ANSWER: holocrine glands

26. Based on their mode of secretion, the salivary glands are classified as:

ANSWER: merocirne glands

27. Based on its mode of secretion, the mammary gland is considered as a:

ANSWER: apocrine gland

28. The most abundant and widespread tissue of the body is:

ANSWER: connective tissue

29. Almost every organ in the body has a connective tissue supporting framework called the:

ANSWER: stroma

30. The epithelial components of an organ are referred to as its:

ANSWER: parenchyma

31. Unlike cells of epithelial tissue, those of connective tissue are separated by large amounts of:

ANSWER: intercellular substanmce

32. Connective tissue develop from an embryonic tissue called:

ANSWER: mesenchyme

33. Inactive fibroblasts are called:

ANSWER: fibrocytes

34. All macrophages and related cells are often referred to collectively as the:

ANSWER: reticuloendothelial system

35. The most numerous of the connective tissue fibers is:

ANSWER: collagen

36. Connective tissue fibers having a yellowish appearance are:

ANSWER: elastic fibers

37. Loose connective tissue is also referred to as:

ANSWER: areolar connective tissue

38. Dense connective tissue may be classified as irregular or:

ANSWER: regular

39. The tiny cavities in which chondrocytes reside are called:

ANSWER: lacunae

40. The dense connective tissue membrane surrounding cartilage is called the:

ANSWER: perichondrium

INSTRUCTIONS: TRUE - FALSE

41. (T/F): Elastic cartilage is found in the external ear.

ANSWER: T

42. (T/F): The lacunae of bone are linked together by a system of tiny tubes called the canaliculi.

ANSWER: T

43. (T/F): Cells located within the lacunae of bone are called osteocytes.

ANSWER: T

44. (T/F): The most common type of cartilage is called fibrocartilage.

ANSWER: F

45. (T/F): The rarest form of cartilage is hyaline cartilage.

ANSWER: F

46. (T/F): Blood and lymph are classified as connective tissue.

ANSWER: T

47. (T/F): Red bone marrow is classified as a connective tissue.

ANSWER: T

48. (T/F): Muscle tissues may be classified as skeletal, cardiac, and smooth.

ANSWER: T

49. (T/F): Generally, cardiac muscle cells contain only one nucleus.

ANSWER: T

50. (T/F): Cell junctions in cardiac muscle are called intercalated discs.

ANSWER: T

51. (T/F): Generally, smooth mucle cells are considered as being multinucleated.

ANSWER: F

52. (T/F): Glial cells are the supporting cells of nervous tissue.

ANSWER: T

53. (T/F): Body cavities that open to the outside are lined by a mucosa, or mucous membrane.

ANSWER: T

54. (T/F): A body cavity that does not open to the outside is lined by a mucous membrane called a mucosa.

ANSWER: F

55. (T/F): An abnormal growth of tissue is called a neoplasm or tumor.

ANSWER: T

56. (T/F): Neoplasms that develop from connective tissues or muscles are referred to as carcinomas.

ANSWER: F

57. (T/F): Neoplasms that develop from epithelial tissues are called sarcomas.

ANSWER: F

58. (T/F): Most cancer cells have one or more mutated genes called oncogenes.

ANSWER: T

59. (T/F): A permanent chemical change in a gene is called a mutation.

ANSWER: T

60. (T/F): The migration of cancer cells from one part of the body to another is called metastasis.

ANSWER: T

INSTRUCTIONS: ANSWER EACH QUESTION ACCORDING TO THE FOLLOWING KEY:

    (A) Only 1 is correct
    (B) Only 2 is correct
    (C) Both are correct
    (D) Neither are correct

61. Metastasis may occur via the:

    1. blood
    2. lymph

ANSWER: C

62. Environmental agents linked to cancer include:

    1. viruses
    2. gamma rays

ANSWER: C

63. Environmental agents linked to cancer:

    1. benzene
    2. cigarette smoke

ANSWER: C

64. A tumor (neoplasm) may be:

   1. malignant
   2. benign

ANSWER: C

65. Red bone marrow:

   1. site of blood cell production
   2. rich in reticular fibers

ANSWER: C

66. Muscle cells may:

   1. be multinucleated
   2. contain only one nucleus per cell

ANSWER: C

67. Intercalated discs:

   1. are characteristic of smooth muscle
   2. occur with great frequency in dense regular connective tissue

ANSWER: D

68. Lacunáe are found in:

   1. bone
   2. cartilage

ANSWER: C

69. Elastic cartilage contains:

   1. collagen
   2. reticular fibers

ANSWER: A

70. Haversian canals are found in:

   1. cartilage
   2. compact bone

ANSWER: B

71. Hyaline cartilage:

   1. is translucent
   2. contains collagen

ANSWER: C

72. Adipocytes:

   1. are found in adipose tissue
   2. are also called fat cells

ANSWER: C

73. Loose connective tissue:

   1. contains fibroblasts
   2. has fewer cells than dende connective tissue

ANSWER: A

74. Mast cells:

   1. are often located along blood vessels
   2. contain heparin and histamine

ANSWER: C

75. An exocrine gland may be:

   1. coiled tubular
   2. acinar

ANSWER: C

76. A simple exocrine gland may be:

   1. tubular
   2. tubuloacinar

ANSWER: A

77. Mesenchyme:

   1. is an embryonic tissue
   2. gives rise to connective tissue

ANSWER: C

78. Transitional epithelium:

    1. is not designed for stretching
    2. is characteristic of the digestive system

ANSWER: D

79. Simple columnar epithelium:

    1. lines the urinary bladder
    2. may be ciliated

ANSWER: B

80. Stratified squamous epithelium:

    1. functions mainly in secretion
    2. lines the upper respiratory passages

ANSWER: D

TB.5:1

TEST BANK for CHAPTER 5: THE INTEGUMENTARY SYSTEM

INSTRUCTIONS: ANSWER EACH QUESTION ACCORDING TO THE FOLLOWING KEY:

> (A) Only 1 is correct
> (B) Only 2 is correct
> (C) Both are correct
> (D) Neither are correct

1. Functions of the skin include:

    1. protection
    2. defense

ANSWER: C

2. Functions of skin include:

    1. stimulus reception
    2. vitamin D synthesis

ANSWER: C

3. Functions of skin include:

    1. waste excretion
    2. maitenance of body temperature

ANSWER: C

4. The skin:

    1. consists, in part, of an outer dermis
    2 is composed, in part, of a thick epidermal layer

ANSWER: D

5. The subcutaneous layer of the skin:

    1. represents a deep extension of the epidermis
    2. serves to anchor skin to underlying tissues

ANSWER: B

6. The epidermis:

    1. is composed of pseudostratified squamous epithelium
    2. is the inner layer of the skin

ANSWER: D

7. The epidermis:

    1. is thin relative to the dermis
    2. may become keratinized

ANSWER: C

8. The epidermis:

    1. consists of several strata
    2. conatins keratinocytes

ANSWER: C

9. Melanocytes:

    1. produce a hormone called melanin
    2. are found in the epidermis

ANSWER: B

10. Cell found in the epidermis:

    1. Langerhans cell
    2. adipocyte

ANSWER: A

11. Cell found in epidermis:

    1. keratinocyte
    2. melanocyte

ANSWER: C

12. The stratum spinosum:

    1. is also called the prickle cell layer
    2. contains cells pushed up from the stratum basale

ANSWER: C

13. The stratum granulosum:

    1. is also called the prickle cell layer
    2. has cells that contain keratohyaline

ANSWER: B

14. Stratum basale is also referred to as the:

    1. stratum germinativum
    2. malpighian layer

ANSWER: C

15. The papillary layer:

    1. is the upper portion of the epidermis
    2. contains an extensive network of capillaries

ANSWER: B

16. The papillary layer:

    1. is the upper portion of the dermis
    2. is structured to help supply oxygen and nutrients to the overlying epidermis

ANSWER: C

17. The reticular layer:

    1. lies deep to the papillary layer
    2. is found within the dermis

ANSWER: C

18. The reticular layer:

    1. is characteristic of the epidermis
    2. lies superficial to the papillary layer

ANSWER: D

19. The superficial fascia:

    1. is a deep extension of the dermis
    2. is also called the subcutaneous layer of the dermis

ANSWER: C

20. A hair follicle:

    1. the root together with its epithelial
       and connective tissue coverings
    2. receives its nourishment from capillaries
       in the papilla

ANSWER: C

21. Hair:

    1. grows continuously, not in cycles
    2. contains keratin

ANSWER: B

22. Genes may code for:

    1. hair color
    2. curly versus straight hair

ANSWER: C

23. Develop from epidermal cells:

    1. hair follicles
    2. sebaceous glands

ANSWER: C

24. Sebaceous glands:

    1. are holocrine glands
    2. sometimes referred to as "oil glands"

ANSWER: C

25. Sebaceous glands:

    1. are merocrine glands
    2. produce sebum

ANSWER: B

26. The substance called sebum:

   1. lubricates the surface of the skin
   2. retards water loss

ANSWER: C

27. The substance called sebum:

   1. contains cholesterol
   2. is produced by sebaceous glands

ANSWER: C

28. Sebum:

   1. may inhibit the growth of certain bacteria
   2. may have antifungal properties

ANSWER: C

29. Sebaceous glands:

   1. are always associated with hairs
   2. are endocrine glands

ANSWER: D

30. Sweat glands are classified as:

   1. eccrine sweat glands
   2. apocrine sweat glands

ANSWER: C

INSTRUCTIONS: TRUE - FALSE

31. (T/F): A blackhead is also referred to as an open comedo.

ANSWER: T

32. (T/F): The black color of an open comedo is due to the presence of dirt.

ANSWER: F

33. (T/F): An open comedo is thought to evolve from closed comedo.
ANSWER: T

34. (T/F): Nails are derivatives of the dermis.
ANSWER: F

35. (T/F): The nail bed lacks a stratum corneum.
ANSWER: T

36. (T/F): The lunula is the white, crescent-shaped region near the root of the nail.
ANSWER: T

37. (T/F): Apocrine sweat glands are found associated with hairs.
ANSWER: T

38. (T/F): Eccrine sweat glands typically discharge their secretions into hair follicles.
ANSWER: F

39. (T/F): Eccrine sweat glands typically discharge their secretions onto the surface of the skin.
ANSWER: T

40. (T/F): Ceruminous glands are modified sweat glands located in the middle ear region.
ANSWER: F

41. (T/F): Cerumen is formed by in part by sebaceous glands.
ANSWER: T

42. (T/F): Warts may be caused by viruses.
ANSWER: T

43. (T/F): Shingles, formally called herpes zoster, is a viral disease characterized buy the presence of painful blisters along certain nerve pathways.

ANSWER: T

44. (T/F): Another term for a condition known as prickly heat is miliaria.

ANSWER: T

45. (T/F): A painful cluster of boils in the deep epidermis is referred to as a callus.

ANSWER: F

46. (T/F): Skin color is in part determied by the presence of oxyhemoglobin.

ANSWER: T

47. (T/F): Skin color is in part determined by the presence of a substance called carotene.

ANSWER: T

48. (T/F): Melanocytes synthesize the pigment melanin from the amino acid tyrosine.

ANSWER: T

49. (T/F): Melanin is a skin pigment that absorbs ultraviolet rays.

ANSWER: T

50. (T/F): Inflammation is characterized in part by pain and heat.

ANSWER: T

51. (T/F): Inflammation is characterized in part by swelling and redness.

ANSWER: T

52. (T/F): Granulation tissue is always associated with wound healing.

ANSWER: F

53. (T/F): Keloids are large bulging scars that extend far beyond the bounds of a wound.

ANSWER: T

54. (T/F): Hypertrophic scars contain excess amounts of elastic fibers.

ANSWER: F

55. (T/F): The most common type of skin cancer is basal cell carcinoma.

ANSWER: T

INSTRUCTIONS: COMPLETE THE FOLLOWING:

56. The most common form of skin cancer is called:

ANSWER: basal cell carcinoma

57. In the process of wound healing, granulation tissue is typically replaced by:

ANSWER: scar tissue

58. Sunburn is typically classified as what type of burn?

ANSWER: first-degree

59. In the skin, the pigment, carotene, is converted to:

ANSWER: vitamin A

60. Melanin is synthesized from the amino acid:

ANSWER: tyrosine

61. The hormone synthesized by the anterior pituitary and that is responsible for stimulating melanin synthesis is:

ANSWER: MSH (melanocyte-stimulating hormone)

62. Another term for the cuticle of the nail is the:

ANSWER: eponychium

63. Ceruminous glands are modified:

ANSWER: sweat glands

64. Sweat glands are classified as being eccrine or:

ANSWER: apocrine

65. Muscles that make one's hairs stand up straight are the:

ANSWER: arrector pili

66. Sebaceous glands secrete a substance called:

ANSWER: sebum

67. The deepest portion of the epidermis is called the:

ANSWER: stratum basale

68. The layer of the epidermis that contains a large quantity of keratohyaline is the:

ANSWER: stratum granulosum

69. The upper region of the dermis is called the:

ANSWER: papillary layer

70. The layer of the epidermis present only in the thick skin of the palms and the soles is called the:

ANSWER: stratum lucidum

71. The amount of time it takes for a basal cell to reach the stratum corneum is approximately:

ANSWER: two weeks

72. The outermost layer of the skin is referred to as the:

ANSWER: epidermis

73. The inner layer of the skin is referred to as the:

ANSWER: dermis

74. The epidermis consists of four or five sublayers or:

ANSWER: strata

75. The most numerous cells of the epidermis are the:

ANSWER: keratinocytes

76. An insoluble protein that imparts strength and waterproofing to skin is:

ANSWER: keratin

77. An epidermal cell type that is produced in the bone marrow is the:

ANSWER: Langerhans cell

78. In the skin a cholesterol compound is converted to:

ANSWER: vitamin D

79. Redness, swelling, heat and pain are the hallmarks of:

ANSWER: inflammation

80. The epithelium beneath the lunula is called the:

ANSWER: nail matrix

TEST BANK for CHAPTER 6: SKELETAL TISSUES

INSTRUCTIONS: SELECT THE ONE BEST ANSWER

1. The wall of the diaphysis is composed of:

a) spongy bone

b) compact bone

c) hyaline cartilage

d) membranous bone

ANSWER: b

2. Epihyseal plates of long bones:

a) are composed of fibrocartilage

b) are located in the middle of the diaphysis

c) function in bone growth

d) do not contain cells capable of mitosis

ANSWER: c

3. The endosteum:

a) lines the walls of the marrow cavity

b) contains more connective tissue fibers than the periosteum

c) does not possess osteogenic potential

d) is characterized by the presence of Sharpey's fibers

4. The diaphysis:

a) is located between the epiphysis and metaphysis

b) represents one end of a long bone

c) is a common feature of a flat bone

d) is also called the shaft of a long bone

ANSWER: d

5. Osteogenic potential is an attribute:

a) of cells called osteoclasts

b) restricted to cells of the periosteum

c) meaning the capacity to destroy bone

d) of certain cells in the endosteum

ANSWER: d

6. Growth hormone indirectly promotes the growth of cartilage by stimulating the liver to synthesize low molecular weight proteins called:

a) glycosaminoglycans

b) low-density lipoproteins

c) somatomedins

d) estrogens

ANSWER: c

7. The vitamin supplement administered in the treatment of rickets is:

a) vitamin A

b) vitamin C

c) vitamin D

d) vitamin K

ANSWER: c

8. Osteitis deformans is also called:

a) vitamin D-resistant rickets

b) osteomyelitis

c) osteoporosis

d) Paget's disease

ANSWER: d

9. A fracture which occurs in a bone that is already diseased is termed a:

a) compound fracture

b) incomplete or green stick fracture

c) pathologic fracture

d) comminuted fracture

ANSWER: c

10. Chondrocytes undergo hypertrophy in the zone of:

   a) maturation
   b) reserve cartilage
   c) proliferation
   d) calcification

ANSWER: a

INSTRUCTIONS: TRUE - FALSE

11. (T/F): Volkmann's canals serve as passageways through which the blood vessels of one osteon interconnect with those of another.

ANSWER: T

12. (T/F): The cartilage model of a long bone grows in length at its ends by the process of appositional growth.

ANSWER: F

13. (T/F): The spaces within which osteoblasts become entrapped are called lamellae.

ANSWER: F

14. (T/F): Osteocytes are stimulated to perform osteocytic osteolysis by the action of parathyroid hormone.

ANSWER: T

15. (T/F): The parietal bones develop by intramembranous bone formation.

ANSWER: T

16. (T/F): One of the earliest events in intramembranous bone formation is the condensation of embryonic mesenchyme.

ANSWER: T

17. (T/F): Most of the bones in the human body are formed by intramembranmous bone formation.

ANSWER: F

18. When ossification is complete only a thin layer of hyaline cartilage remains on the surface of the epiphysis and is known as articular cartilage.

ANSWER: T

19. Microcopic examination reveals that the epiphyseal plate is composed of seven distinct regions.

ANSWER: F

20. The zone of calcification serves to unite the epiphyseal plate with the diaphyseal bone.

ANSWER: T

INSTRUCTIONS: COMPLETE THE FOLLOWING STATEMENTS.

21. Tendons having a broad, flattened appearance such as those of the external and internal abdominal oblique muscles are called:

ANSWER: aponeuroses

22. The youngest chondrocytes within the zone of maturation are those located closest to the zone of:

ANSWER: proliferation

23. Large, multinucleated cells that function in the resorption of bone are called:

ANSWER: osteoclasts

24. Osteoblasts deposit collagen and ground substance around themselves and in doing so become trapped within spaces called:

ANSWER: lacunae

25. The ability of certain cells to form bone is called:

ANSWER: osteogenic potential

26. A low level of calcium in the blood is called:

ANSWER: hypocalcemia

27. The soluble protein synthesized by osteoblasts which, when exported from the cell becomes collagen and forms fibers spontaneously is called:

ANSWER: tropocollagen

28. Osteoblasts become osteocytes once the surrounding matrix becomes:

ANSWER: calcified

29. One of the functions of skeletal tissue is the formation of blood cells or:

ANSWER: hematopoiesis

30. A bursa is an anatomical structure which serves to reduce:

ANSWER: frictional stress

INSTRUCTIONS: SELECT THE MOST APPROPRIATE ANSWER FROM COLUM B. THE SAME ANSWER MAY BE USED MORE THAN ONCE OR NOT AT ALL.

Column A	Column B
31. ( ) Osteogenesis imperfecta	A. may be the result of a serious bacterial or fungal infection
32. ( ) Osteomalacia	B. may derive from vitamin D deficiency in children
33. ( ) Osteosarcoma	C. net result of this disease is a loss of bone mass
34. ( ) Osteitis deformans	D. second most common form of malignancy in young people
35. ( ) Osteomyelitis	E. Paget's disease

36. ( ) Rickets   F. literally it means a
                                              "softening" of bone

37. ( ) Vitamin D-resistant   G. may result from a
         rickets   condition known as
                                              congenital
                                              hypophosphatemia

38. ( ) Osteoporosis   H. patient may also have
                                              discoloration of teeth,
                                              deafness, and sclerae
                                              with a bluish tint

ANSWERS: 31.(H), 32.(F), 33.(D), 34.(E),
         35.(A), 36.(B), 37.(G), 38.(C)

INSTRUCTIONS: ANSWER ACCORDING TO THE FOLLOWING KEY:

(A) only 1,2 and 3 are CORRECT
(B) only 1 and 3 are CORRECT
(C) only 2 and 4 are CORRECT
(D) only 4 is CORRECT
(E) ALL are CORRECT

39. Functions provided by the skeletal system include:

1. support
2. movement
3. hematopoiesis
4. protection

ANSWER: A

40. Parts of a typical long bone include the:

    1. metaphysis
    2. peritendinium
    3. epiphysis
    4. perichondrium

ANSWER: B

41. Sharpey's fibers:

    1. are composed entirely of elastic fibers
    2. help stabilize the articular cartilage
    3. are located primarily in the endosteum
    4. anchor periosteum to underlying bone

ANSWER: D

42. Cells with osteogenic potential:

    1. are located within the endosteum
    2. exhibit most of their activity following closure of the epiphyseal plates
    3. are located within the inner layer of the periosteum
    4. actively participate in bone resorption

ANSWER: B

43. During fracture healing:

   1. a fibrocartilaginous callus forms prior to a procallus
   2. clotted blood is converted to a procallus
   3. the osseous callus typically attains its maximum size in 5 or 6 days
   4. new bone formation is observable within the first week

ANSWER: C

44. Secondary centers of ossification:

   1. are located in the epiphyses of long bones
   2. develop as calcified cartilage breaks down
   3. are invaded by blood vessels as they develop
   4. are invaded by osteoblasts as they develop

ANSWER: E

45. Components of the extracellular matrix of bone include:

   1. hydroxyapatite crystals
   2. collagen fibers
   3. carbonate and citrate ions
   4. osteoid

ANSWER: E

46. The organic matrix of bone:

    1. represents approximately 50% of its dry weight
    2. is partially composed of ground substance
    3. contains glycosaminoglycans
    4. consists largely of calcium phosphate

ANSWER: A

47. Type I collagen fibers:

    1. are secreted by the osteoblasts
    2. represent about 90% of the organic matrix of bone
    3. each consist of five intertwined strands of protein
    4. are characteristic of the matrix of bone

ANSWER: C

48. Osteoprogenitor cells:

    1. differentiate to become osteoblasts
    2. differentiate to become cells capable of producing the matrix of bone
    3. are active during fracture healing
    4. reside within the firbous periosteum

ANSWER: A

49. Osteoclasts:

   1. are active in bone remodeling
   2. function in the removal of pre-existing bone
   3. may be derived from fusion white blood cells
   4. are activated by elevated levels of thyroid hormone in the blood

ANSWER: A

50. Osteocytes:

   1. differentiate from osteoblasts
   2. are located within lacunae
   3. establish contact with one another through canaliculi
   4. respond to parathyroid hormone by releasing calcium from bone

ANSWER: E

INSTRUCTIONS: ANSWER EACH QUESTION ACCORDING TO THE FOLLOWING KEY:

(A) Only 1 is correct
(B) Only 2 is correct
(C) Both are correct
(D) Neither are correct

51. The epiphyseal plate of a long bone:

    1. plays an important role in bone growth
    2. is located in the center of the diaphysis

ANSWER: A

52. The epiphyseal plate of a long bone:

    1. typically lies parallel to the long axis of the bone
    2. are not usually replaced by bone until the bone has attained its final length

ANSWER: B

53. The zone of resting or reserve cartilage:

    1. lies adjacent to the zone of hypertrophy
    2. consists primarily of fibrocartilage

ANSWER: D

54. Lengthening of the diaphysis of a long bone:

   1. usually occurs as a result of interstitial growth
   2. terminates with closure of the epiphyseal plates

ANSWER: B

55. Appositional growth:

   1. involves the removal of pre-existing bone
   2. is a process by which the diaphysis increases in diameter

ANSWER: C

56. Woven bone:

   1. represents an early stage of intramembranous bone formation
   2. is characterized by the orderly arrangement of collagen fibers

ANSWER: A

57. Spongy bone:

   1. is typically found in the epiphyses of long bones
   2. is also referred to as cancellous bone

ANSWER: C

58. The marrow cavity:

   1. is typically occupied by compact bone
   2. does not increase in diameter during bone growth

ANSWER: D

59. The osteogenic layer of the periosteum:

   1. is located deep to the fibrous periosteum
   2. is restricted to the region of the metaphysis

ANSWER: A

60. Certain radioactive isotopes:

   1. can have harmful effects on bone and its hematopoietic capacity
   2. serve as valuable tools in investigations of normal and abnormal bone physiology

ANSWER: C

61. A tendon:

   1. is composed largely of loose connective tissue
   2. contains elastic fibers which are responsible for its shiny, white appearance

ANSWER: D

62. Structures which prevent or reduce frictional stress and or damage to tendons include:

   1. bursae
   2. tendon sheaths

ANSWER: C

63. Ligaments:

   1. serve to bind bones or bony parts together
   2. are composed of collagen and elastic fibers

ANSWER: C

64. Ligaments:

   1. function primarily to limit the range of motion of joints
   2. possess a rich vascular supply and therefore heal rapidly following injury

ANSWER: A

65. The collateral ligaments of the knee joint:

   1. actually represent portions of the fibrous capsule of the joint itself
   2. function primarily to reduce frictional stress

ANSWER: A

66. Steatorrhea is a condition which:

   1. may result in a vitamin D deficiency
   2. is characterized by the inability to absorb fat from the gastrointestinal tract

ANSWER: C

67. Somatomedins are low molecular weight proteins which are synthesized by cells in the:

   1. periosteum
   2. endosteum

ANSWER: D

68. Vitamin D:

   1. belongs to the group of water soluble vitamins
   2. is lost in the feces of patients having a condition called steatorrhea

ANSWER: B

69. Calcitonin is a hormone which:

   1. elevates the level of calcium in the blood
   2. stimulates osteoclastic activity

ANSWER: D

70. Hormone responsible for elevating the level of calcium in the blood:

   1. parathyroid hormone
   2. calcitonin

ANSWER: A

71. Parathyroid hormone:

   1. lowers the level of calcium in the blood
   2. inhibits the excretion of calcium by the kidneys

ANSWER: B

72. Its prolonged secretion promotes the closure of epiphyseal plates of long bones:

   1. estrogen
   2. testosterone

ANSWER: C

73. Lamellar bone:

   1. is typically deposited in concentric rings around a centrally located blood vessel during remodeling
   2. is characterized by the presence of osteons or haversian systems

ANSWER: C

74. Canaliculi:

   1. serve to connect osteocytes within a given osteon
   2. provide a channel by which osteocytes obtain nutrients and rid themselves of metabolic waste products

ANSWER: C

75. Remodeling of bone:

   1. occurs only during periods of active bone growth
   2. involves osteoclastic activity

ANSWER: B

76. Researchers have demonstrated that weak electrical fields:

   1. inhibit the processes of bone repair and remodelling
   2. may have a potential use in the repositioning of teeth

ANSWER: B

77. Woven bone is:

   1. uniquely suitable for the stresses and strains encountered by the adult skeleton
   2. formed only during the process of endochondral bone development

ANSWER: D

78. The closure of epiphyseal plates of long bones:

   1. generally occurs at an earlier age in males
   2. may be detected by radiographic (x-ray) assessment

ANSWER: B

79. Cells contributing to the growth in length of a long bone:

   1. chondrocytes in the zone of resting or reserve cartilage
   2. chondrocytes in the zone of proliferation

ANSWER: B

80. Bones typically developing by the process of intramembranous bone formation:

   1. several bones of the skull
   2. long bones of the limbs

ANSWER: A

ESSAY QUESTIONS

1. Describe the role of cell death in bone growth occurring at the epiphyseal plate.

2. Explain the biological significance of canaliculi and Volkmann's canals as they relate to bone physiology.

3. Describe the roles of the thyroid and parathyroid glands in bone physiology.

4. Explain how growth of long bones in the female typically ceases prior to that in males.

TEST BANK for CHAPTER 7: THE AXIAL SKELETON

INSTRUCTIONS: SELECT THE ONE BEST ANSWER

1. The sternum may be classified as a (an):

a) long bone
b) short bone
c) flat bone
d) irregular bone

ANSWER: c

2. The scapula may be classified as a (an):

a) long bone
b) short bone
c) flat bone
d) irregular bone

ANSWER: c

3. Which is not typically classified as a long bone?

a) humerus
b) radius
c) tibia
d) talus

ANSWER: d

4. A hook-like projection on the surface of a bone is called a:

a) process
b) facet
c) hamulus
d) condyle

ANSWER: c

5. An articular surface that is shaped like a pulley is called a:

a) facet
b) meatus
c) sulcus
d) trochlea

ANSWER: d

6. An elongated depression on the surface of a bone is called a:

a) fossa
b) sulcus
c) meatus
d) foramen

ANSWER: b

7. Wormian bones are also referred to as:

a) sesamoid bones
b) sutural bones
c) irregular bones
d) short bones

ANSWER: b

8. The number of bones comprising the axial skeleton is:

a) 40
b) 60
c) 80
d) 100

ANSWER: c

9. The number of bones comprising the skull is:

a) 22
b) 24
c) 26
d) 30

ANSWER: a

10. The number of ribs in the human body is:

a) 20
b) 22
c) 24
d) 26

ANSWER: c

11. The bone of the skull that contains ear ossicles is the:

a) temporal
b) parietal
c) frontal
d) sphenoid

ANSWER: a

12. The number of bones comprising the cranium is:

a) 4
b) 6
c) 8
d) 10

ANSWER: c

13. Which is not a bone of the cranium?

a) ethmoid
b) occipital
c) parietal
d) lacrimal

ANSWER: d

14. Which is not a bone of the facial skeleton?

a) zygoma
b) malar
c) temporal
d) vomer

ANSWER: c

15. Which is not a bone of the axial skeleton?

a) sternum
b) clavicle
c) lacrimal
d) sacrum

ANSWER: b

16. Which is not a characteristic feature of the mandible?

a) mental foramen
b) coronoid process
c) condyloid process
d) styloid process

ANSWER: d

17. Which is not a characteristic feature of the temporal bone?

a) external auditory meatus
b) mastoid process
c) lacrimal fossa
d) styloid process

ANSWER: c

18. Which bone does not form part of the middle cranial fossa?

a) frontal
b) sphenoid
c) parietal
d) temporal

ANSWER: a

19. The foramen magnum is a characteristic feature of which bone?

a) frontal
b) parietal
c) temporal
d) occipital

ANSWER: d

20. The median nuchal crest is found on which bone?

a) occipital
b) temporal
c) frontal
d) sphenoid

ANSWER: a

21. The supraorbital notch or foramen is a chacteristic feature of which bone?

a) occipital
b) temporal
c) frontal
d) nasal

ANSWER: c

22. The glabella is located on which bone?

a) frontal
b) zygoma
c) nasal
d) temporal

ANSWER: a

23. Which bone does not contain a paranasal sinus?

a) occipital
b) sphenoid
c) ethmoid
d) maxilla

ANSWER: a

24. The hypoglossal canals are characteristic features of which bone?

a) temporal
b) sphenoid
c) occipital
d) frontal

ANSWER: c

25. The optic canals are characteristic features of which bone?

a) frontal
b) sphenoid
c) occipital
d) ethmoid

ANSWER: b

26. The cribriform plate is a characteristic feature of which bone?

a) frontal
b) sphenoid
c) maxialla
d) ethmoid

ANSWER: d

27. The sella turcica is a characteristic feature of which bone?

a) ethmoid
b) sphenoid
c) vomer
d) palatine

ANSWER: b

28. The clivus is a characteristic feature of which bone?

a) occipital
b) ethmoid
c) vomer
d) sphenoid

ANSWER: d

29. The carotid canal is a characteristic feature of which bone?

a) sphenoid
b) ethmoid
c) temporal
d) parietal

ANSWER: c

30. Cranial nerve which does not pass through the jugular foramen?

a) vagus (CN X)
b) spinal accessory (CN XI)
c) glossopharyngeal (CN IX)
d) facial (CN VII)

ANSWER: d

31. The stylomastoid foramen is a characteristic feature of which bone?

a) temporal
b) parietal
c) sphenoid
d) frontal

ANSWER: a

32. The 'lower jaw' is also called the:

a) maxilla
b) mandible
c) zygoma
d) vomer

ANSWER: b

33. The infraorbital foramen is a characteristic feature of which bone?

a) zygoma
b) frontal
c) maxilla
d) mandible

ANSWER: c

34. Which is not a characteristic feature of the maxilla?

a) frontal process
b) zygomatic process
c) alveolar process
d) styloid process

ANSWER: d

35. The number of cervical vertebrae is:

a) 6
b) 7
c) 8
d) 9

ANSWER: b

36. The number of throacic vertebrae is:

a) 8
b) 10
c) 12
d) 14

ANSWER: c

37. Which is not a charcteristic feature of the axis?

a) dens
b) odontoid process
c) transverse foramen
d) demifacet

ANSWER: d

38. Which cervical vertebra is called the vertebra prominens?

a) first
b) fourth
c) sixth
d) seventh

ANSWER: d

39. How many pairs of true or vertebrosternal ribs are there in the thoracic skeleton?

a) 5
b) 6
c) 7
d) 8

ANSWER: c

40. How many pairs of false ribs are there in the thoracic skeleton?

a) 5
b) 6
c) 7
d) 8

ANSWER: a

41. How many pairs of ribs articulate directly with the sternum via their costal cartilages?

a) 5
b) 6
c) 7
d) 8

ANSWER: a

42. How many ribs articulate with the manubrium of the sternum?

a) 2
b) 4
c) 6
d) 8

ANSWER: a

43. How many ribs articulate with the body of the sternum?

a) 6
b) 8
c) 10
d) 12

ANSWER: c

44. The most superior portion of the sternum is called the:

a) xiphoid process
b) body
c) manubrium
d) styloid process

ANSWER: c

45. The typical number of lumbar vertebrae is:

a) 3
b) 4
c) 5
d) 6

ANSWER: c

46. Which is not a characteristic feature of a typical vertebra?

a) vertebral canal
b) neural arch
c) transverse foramen
d) centrum

ANSWER: c

47. Which cervical vertebra is also called the axis?

a) first
b) second
c) third
d) seventh

ANSWER: b

48. Which bone does not articulate with another bone?

a) vomer
b) hyoid
c) lacrimal
d) ethmoid

ANSWER: b

49. Which is not a characteristic feature of the sphenoid bone?

a) foramen rotundum
b) foramen ovale
c) pterygoid canal
d) carotid canal

ANSWER: d

50. Which is not an unpaired bone?

a) parietal
b) ethmoid
c) vomer
d) hyoid

ANSWER: a

INSTRUCTIONS: TRUE - FALSE

51. (T/F): The squamosal suture is a joint between the temporal and parietal bones of the skull.

ANSWER: T

52. (T/F): The three bones which form the foramen lacerum are the sphenoid, temporal and occipital.

ANSWER: T

53. (T/F): The greater and lesser wings are prominent features of the temporal bone.

ANSWER: F

54. (T/F): The cribriform plate is a characteristic feature of the anterior cranial fossa.

ANSWER: T

55. (T/F): The superior orbital fissure is situated between the greater and lesser wings of the sphenoid bone.

ANSWER: T

56. (T/F): Perforations in the cribriform plate transmit fibers of the olfactory nerve (CN I).

ANSWER: T

57. (T/F): The foramen spinosum transmitts fibers of the facial nerve (CN VII).

ANSWER: F

58. (T/F): The nerve of the pterygoid canal passes through a portion of the ethmoid bone.

ANSWER: F

59. (T/F): The internal carotid artery enters the middle cranial fossa by passing through the foramen lacerum.

ANSWER: F

60. (T/F): The anterior and posterior ethmoidal foramina are located along the junction of the medial wall and roof of the orbit.

ANSWER: T

61. (T/F): Fibers of the maxillary division of the trigeminal nerve (CN V) pass through the foramen rotundum.

ANSWER: T

62. (T/F): The Foramen rotundum is a characteristic feature of the anterior cranial fossa.

ANSWER: F

63. (T/F): The foramen spinosum is located in the posterior cranial fossa.

ANSWER: F

64. (T/F): The incisive foramen transmitts the nasopalatine nerve.

ANSWER: T

65. (T/F): The occipital condyles may be observed on the inferior aspect of the base of the skull.

ANSWER: T

66. (T/F): The periosteal lining on the external surface of the skull is called the endocranium.

ANSWER: F

67. (T/F): The Pericranium is the periosteal covering located on the internal aspect of the skull.

ANSWER: F

68. (T/F): The sagittal suture occupies a position between the frontal and temporal bones.

ANSWER: F

69. (T/F): The lambdoid suture occupies a position between the occipital and parietal bones.

ANSWER: T

70. (T/F): The cribriform plate is a characteristic feature of the onferior aspect of the ethmoid bone.

ANSWER: F

71. (T/F): The cerebellum occupies a position within the posterior cranial fossa.

ANSWER: T

72. (T/F): The vomer contributes to the formation of the nasal septum.

ANSWER: T

73. (T/F): The vomer articulates with the maxillae, the palatine bones, and the ethmoid bone.

ANSWER: T

74. (T/F): The superior and middle nasal conchae represent separate bones of the skull.

ANSWER: F

75. (T/F): The maxillae and mandible each possess an alveolar process.

ANSWER: T

76. (T/F): The single sacrum arises through the fusion of five individual vertebrae during development.

ANSWER: T

77. (T/F): Funnel chest refers to a condition in which the sternum lies below the level of the chest wall.

ANSWER: T

78. (T/F): The external connective tissue fibers of an intervertebral disc are collectively known as the nucleus pulposus.

ANSWER: F

79. (T/F): Intervertebral discs bear significant gravitational pressures and function somewhat like chusions.

ANSWER: T

80. (T/F): The cervical and lumbar curves of the vertebral column are also referred to as secondary curves.

ANSWER: T

81. (T/F): The thoracic and lumbar curves of the vertebral column are designated as being primary curves.

ANSWER: F

INSTRUCTIONS: COMPLETE THE FOLLOWING STATEMENTS.

82. The third rib articulates with the lateral aspect of the sternum via its:

ANSWER: costal cartilage

83. Transverse foramina are a characteristic feature of which type of vertebrae?

ANSWER: cervical

84. The dens or odontoid process is a characteristic feature of which cervical vertebra?

ANSWER: second (axis)

85. The dens articulates with the anterior arch of the:

ANSWER: atlas (first cervical vertebra)

86. The palatine bones contribute to the formation of the hard:

ANSWER: palate

87. The lacrimal sac is typically housed within a depression on the lacrimal bone termed the lacrimal:

ANSWER: fossa

88. A groove for the subclavian artery and vein represent characteristic features of the superior aspect of the first:

ANSWER: rib

89. The constriction located between the head and the tubercle of a rib is called its:

ANSWER: neck

90. The interval bewteen adjacent ribs is called an:

ANSWER: intervertebral space

91. The tenth and eleventh pairs of ribs are typically referred to as:

ANSWER: floating (vertebral) ribs

92. The hyoid bone possesses two prominent projections on each side called the greater and lesser:

ANSWER: cornua

93. Failure of the palatine processes of the maxillae to fuse during development leads to a clinical condition called:

ANSWER: cleft palate

94. The madibular foramen is located adjacent to a bony projection called the:

ANSWER: lingula

95. The sella turcica or 'Turkish saddle' serves to accomodate the:

ANSWER: pituitary gland

96. A prominent midline projection of the ethmoid bone seen in the anterior cranial fossa is called the:

ANSWER: crista galli

97. The temporal lobes of the cerebral hemispheres are located with the middle cranial:

ANSWER: fossa

98. The brainstem above becomes continuous with the spinal cord below through an opening in the occipital bone called the:

ANSWER: foramen magnum

99. The mandibular division of the trigeminal nerve (CN V) is transmitted via an opening called the foramen:

ANSWER: ovale

TEST BANK for CHAPTER 8: APPENDICULAR SKELETON

INSTRUCTIONS: SELECT THE ONE BEST ANSWER

1. Which is not a characteristic feature of the scapula:

a) inferior angle
b) axillary border
c) glenoid fossa
d) conoid tubercle

ANSWER: d

2. Which is not a characteristic feature of the ulna:

a) coronoid process
b) olecranon process
c) trochlea
d) styloid process

ANSWER: c

3. The number of metacarpals in each hand is:

a) 5
b) 6
c) 7
d) 8

ANSWER: a

4. The number of carpal bones in each hand is:

a) 5
b) 6
c) 7
d) 8

ANSWER: d

5. The number of phalanges in the first digit of each hand is:

a) 2
b) 3
c) 4
d) 5

ANSWER: a

6. Which is not a characteristic feature of the distal aspect of the humerus:

a) trochlea
b) olecranon fossa
c) anatomical neck
d) capitulum

ANSWER: c

7. The nerve that occupies a position in close proximity to the medial epicondyle of the humerus is the:

a) ulnar nerve
b) radial nerve
c) median nerve
d) musculocutaneous nerve

ANSWER: a

8. The number of bones in each upper limb is:

a) 20
b) 30
c) 35
d) 40

ANSWER: 30

9. The shaft of the humerus has a characteristic groove for which nerve:

a) radial
b) median
c) ulnar
d) musculocutaneous

ANSWER: a

10. Which cannot be observed from the posterior aspect of the humerus:

a) coronoid fossa
b) olecranon fossa
c) trochlea
d) medial epicondyle

ANSWER: a

INSTRUCTIONS: ANSWER EACH QUESTION ACCORDING TO THE FOLLOWING KEY:

    (A) Only 1 is correct
    (B) Only 2 is correct
    (C) Both are correct
    (D) Neither are correct

11. Characteristic feature of the proximal humerus:

    1. intertubercular groove
    2. lesser tubercle

ANSWER: C

12. The surgical neck of the humerus:

    1. located distal to the anatomical neck
    2. fractures here may injure the axillary nerve

ANSWER: C

13. The medial epicondyle of the humerus:

    1. located superomedial to the trochlea
    2. related posteriorly to the ulnar nerve

ANSWER: C

14. The elongated depression located between the greater and lesser tubercles of the humerus:

    1. bicipital groove
    2. intertubercular groove

ANSWER: C

15. The trochlea of the humerus:

    1. shaped like a pulley
    2. located medial to the capitulum

ANSWER: C

16. The interosseous membrane:

    1. unites the radius and the ulna
    2. unites the tibia and the fibula

ANSWER: C

17. The shoulder girdle:

    1. also called the pectoral girdle
    2. comprised of four bones

ANSWER: A

18. The scapula:

    1. is typically classified as an irregular bone
    2. articulates with the vertebral column

ANSWER: D

19. Each pectoral (shoulder) girdle:

    1. consists of two bones
    2. is formed in part by the humerus

ANSWER: A

20. The glenoid fossa of the scapula:

    1. is best seen in a lateral view of that bone
    2. articulates with the clavicle

ANSWER: A

21. Bone contributing to the formation of the shoulder girdle:

    1. clavicle
    2. scapula

ANSWER: C

22. Characteristic features of the scapula:

    1. conoid tubercle
    2. trapezoid line

ANSWER: D

23. Cleidocranial dysostosis is a condition involving defects of the:

    1. clavicle
    2. skull

ANSWER: C

24. Located in the distal row of carpal bones:

    1. hamate
    2. trapezoid

ANSWER: C

25. Located in the distal row of carpal bones:

    1. pisiform
    2. scaphoid

ANSWER: D

26. Located in the distal row of carpal bones:

    1. trapezium
    2. lunate

ANSWER: A

27. Located in the proximal row of carpal bones:

    1. triquetral
    2. hamate

ANSWER: A

28. Possesses a characteristic prominence called the styloid process:

    1. humerus
    2. ulna

ANSWER: B

29. Possesses a characteristic feature called the styloid process:

    1. radius
    2. ulna

ANSWER: C

30. Forms an articulation with the distal end of the ulna:

    1. ulnar notch
    2. styloid process

ANSWER: A

31. Regarding the metacarpals:

    1. their bases are directed distally
    2. their heads are directed proximally

ANSWER: D

32. 'Knuckles' are formed by the:

    1. bases of the metacarpals
    2. heads of the metacarpals

ANSWER: B

33. The pelvic girdle:

    1. consists of the two coxal or innominate bones
    2. is united anteriorly at the pubic symphysis

ANSWER: A

34. Located inferior to the ischial spine:

    1. greater sciatic notch
    2. lesser sciatic notch

ANSWER: B

35. Serves for articulation with the head of the femur:

    1. obturator foramen
    2. acetabulum

ANSWER: B

36. The skeleton of the pelvis:

    1. consists in part of the sacrum and coccyx
    2. consists of only three bones

ANSWER: A

37. Characteristic feature of the coxal bone:

    1. ischial tuberosity
    2. sacral promontory

ANSWER: A

38. Which is not a characteristic feature of the coxal bone:

    1. obturator foramen
    2. anterior inferior iliac spine

ANSWER: D

39. Articulates with the head of the femur:

    1. lunate surface of acetabulum
    2. acetabular fossa

ANSWER: A

40. The pubic arch:

    1. more acute in females than males
    2. also called the subpubic angle

ANSWER: B

INSTRUCTIONS: TRUE - FALSE

41. (T/F): The ischial spine is located superior to the lesser sciatic notch.

ANSWER: T

42. (T/F): The pelvic inlet is situated superior to the pelvic outlet.

ANSWER: T

43. (T/F): The true pelvis is located inferior to the plane of the superior pelvic aperature.

ANSWER: T

44. (T/F): The greater or false pelvis lies at a level superior to that of the true pelvis.

ANSWER: T

45. (T/F): The skeleton of the pelvis is composed of two bones from the appendicular skeleton and two from the axial skeleton.

ANSWER: T

46. (T/F): The tibia is located lateral to the fibula in the anatomical position.

ANSWER: F

47. (T/F): The linea aspera is located primarily on the anterior aspect of the shaft of the femur.

ANSWER: F

48. (T/F): The linea aspera is a characteristic feature of the tibia.

ANSWER: F

49. (T/F): The base of the patella is directed superiorly.

ANSWER: T

50. (T/F): The intercondylar fossa is best seen from the anterior aspect of the femur.

ANSWER: F

51. (T/F): The lesser trochanter of the femur is located in a posteromedial position relative to the greater trochanter.

ANSWER: T

52. (T/F): The fovea capitis represents a characteristic feature of the proximal tibia.

ANSWER: F

53. (T/F): The angle of inclination is formed between the neck and the shaft of the femur.

ANSWER: T

54. (T/F): The intertrochanteric crest is best seen in an anterior view of the femur.

ANSWER: T

55. (T/F): The medial malleolus is a characteristic feature of the fibula.

ANSWER: F

56. (T/F): Only one bone of the leg articulates with the talus.

ANSWER: F

57. (T/F): The tibial tubercle is best seen in a posterior view of the tibia.

ANSWER: F

58. (T/F): The patellar ligament is attached to the tibial tubercle.

ANSWER: T

59. (T/F): The malleolar groove is located on the posterior aspect of the distal fibula.

ANSWER: F

60. (T/F): The plantar surface of the foot is represented by the sole.

ANSWER: T

COMPLETE THE FOLLOWING STATEMENTS

61. The number of tarsal bones in each foot is:

ANSWER: seven

62. The metatarsal bone located most laterally in the foot is the:

ANSWER: fifth

63. The fifth metatarsal bone articulates with a tarsal bone called the:

ANSWER: cuboid

64. The tibia and fibula articulate with a tarsal bone called the:

ANSWER: talus

65. The proximal extremity of each phalanx is called its:

ANSWER: base

66. The distal extremity of each metatarsal is called its:

ANSWER: head

67. The portion of a metatarsal located between its head and base is called the:

ANSWER: shaft

68. The cuneiform bone that is located most medially in the foot is the:

ANSWER: first

69. The tarsal bone that articulates with all three of the cuneiform bones is called the:

ANSWER: navicular

70. The distal end of the fibula is expended to form the:

ANSWER: lateral malleolus

71. The lateral expanded portion of the scapular spine is called the:

ANSWER: acromion process

72. The medial border of the scapula is also called the:

ANSWER: vertebral border

73. The layman's term for the clavicle is the:

ANSWER: collarbone

74. When we are sitting a large portion of our weight is supported by features of the coxal bones called the:

ANSWER: ischial tuberosities

75. The expanded winglike area of the ilium is also called the:

ANSWER: ala

76. The ischial spine may be palpated per vaginum in order to locate the:

ANSWER: pudendal nerve

77. The inferior rami of the pubic bones forms the:

ANSWER: pubic arch (subpubic angle)

78. The feature of the scapula to which the short head of the biceps and the coracobrachialis muscles attach is called the:

ANSWER: coracoid process

79. The layman's term for the scapula is the:

ANSWER: 'shoulder blade'

80. The bicipital groove is located between the greater and lesser:

ANSWER: tubercles

81. A congenital malformation involving partial or complete absence the scapula together with delayed ossification of certain bones of the skull is called:

ANSWER: cleidocranial dysostosis

TB.9:1

TEST BANK for CHAPTER 9: JOINTS

TEST BANK for CHAPTER 3: THE CELL: BUILDING BLOCK OF THE BODY

INSTRUCTIONS: SELECT THE MOST APPROPRIATE ANSWER FROM COLUMN B. THE SAME ANSWER MAY BE USED ONCE, MORE THAN ONCE, OR NOT AT ALL.

Column A	Column B
1. bursitis	A. a 'peg-and-socket type joint
2. diarthroses	B. classification term for joints considered to be immovable
3. synarthroses	C. inflammation of synovial tendon sheaths
4. synostosis	D. inflammation of a bursa
5. gomphosis	E. classification term for joints considered to be freely movable
6. syndesmosis	F. fluid-filled fibrous sacs located near joints and which serve to reduce frictional forces
7. symphysis	G. a type of cartilaginous joint
8. synchondrosis	H. a type of fibrous joint
9. bursae	I. conversion of fibrous connective tissue within a suture to bone
10. tenosynovitis	J. a disorder of uric acid metabolism

ANSWERS: 1. (D), 2. (E), 3. (B), 4. (I), 5. (A), 6. (H), 7. (G) 8. (G), 9. (F), 10. (C)

INSTRUCTIONS: ANSWER EACH QUESTION ACCORDING TO THE FOLLOWING KEY:

        (A) Only 1 is correct
        (B) Only 2 is correct
        (C) Both are correct
        (D) Neither are correct

11. Joint classified as a gomphosis:

    1. a type of fibrous joint
    2. a 'peg-and-socket type joint

ANSWER: C

12. A type of fibrous joint:

    1. syndesmosis
    2. suture

ANSWER: C

13. A syndesmosis:

    1. is a type of synovial joint
    2. is a slightly movable fibrous joint

ANSWER: B

14. The distal tibiofibular joint:

    1. a type of synovial joint
    2. contains a cartilaginous disc

ANSWER: D

15. The knee joint:

    1. contains intracapsular ligaments
    2. is a synovial joint

ANSWER: C

16. The symphysis pubis:

    1. is a type of fibrous joint
    2. is a synchondrosis

ANSWER: D

17. Synchondroses are:

   1. a special type of cartilaginous joint
   2. of a temorary nature

ANSWER: C

18. Synchondroses:

   1. permit growth of bones to occur
   2. represent a special type of fibrous joint

ANSWER: A

19. Characteristic feature of a synovial joint:

   1. a joint capsule
   2. articular cartilage

ANSWER: C

20. A bursa may:

   1. communicate with the synovial cavity of a joint
   2. serve to reduce frictional forces

ANSWER: C

21. The synovial membrane:

   1. covers the articular surfaces of a synovial joint
   2. produces synovial fluid

ANSWER: B

22. Regarding synovial fluid contains:

   1. macrophages
   2. polymorphonuclear leukocytes

ANSWER: C

23. Regarding synovial fluid:

   1. is typically viscous
   2. contains hyaluronic acid

ANSWER: C

24. Regarding the A cells of the synovial membrane:

   1. produce hyaluronic acid
   2. are phagocytic

ANSWER: C

25. Articular fat pads are thought to:

   1. increase the surface area of the synovial membrane
   2. facilitate the distribution of synovial fluid within the joint

ANSWER: C

26. Regarding accessory ligaments of synovial joints:

   1. may be located in an extracapsular position
   2. are simply thickenings of the fibrous capsule

ANSWER: A

27. Synovial joints may be classified as:

   1. simple
   2. compound

ANSWER: C

28. Synovial joints may be classified as:

   1. complex
   2. simple

ANSWER: C

29. Ginglymus joints are:

   1. typically uniaxial
   2. also called hinge joints

ANSWER: C

30. The talocrural (ankle) joint is a:

   1. ginglymus joint
   2. synovial joint

ANSWER: C

31. The talocrural (ankle) joint represents a:

    1. type of diarthroses
    2. representative ginglymus joint

ANSWER: C

32. The atlantoaxial joint is:

    1. a trochoid type joint
    2. one that permits rotation to occur

ANSWER: C

33. A type of angular movement:

    1. flexion
    2. extension

ANSWER: C

34. Example of a spheroid type synovial joint:

    1. knee
    2. atlantoaxial

ANSWER: D

35. Proposed functions attributable to an articular disc:

    1. shock absorption
    2. joint stability

ANSWER: C

INSTRUCTIONS: TRUE - FALSE

36. (T/F): Synovial joints are most often classified according to their shape.

ANSWER: T

37. (T/F): The anterior and posterior cruciate ligaments of the knee are of the intracapsular type.

ANSWER: T

38. (T/F): The tibial collateral ligament of the knee is an accessory type ligament.

ANSWER: T

39. (T/F): The capsule of the knee joint is strengthened anteriorly by the oblique popliteal ligament.

ANSWER: F

40. (T/F): The acetabular labrum forms a complete ring around the periphery of the acetabulum.

ANSWER: F

41. (T/F): Movement about one, two, or three perpendicular axes is a means for classifying synovial joints.

ANSWER: T

42. (T/F): The intermetatarsal joints are examples of hinge or ginglymus synovial joints.

ANSWER: F

43. (T/F): Some intercarpal and intertarsal joints are classified as synovial joints of the plane or gliding type.

ANSWER: T

44. (T/F): The proximal radioulnar joint is a synovial joint of the trochoid or pivot type.

ANSWER: T

45. (T/F): Ellipsoid joints are biaxial.

ANSWER: T

46. (T/F): Spheroid synovial joints are multiaxial.

ANSWER: T

47. (T/F): The carpometacarpal joint of the thumb is a synovial joint of the saddle type.

ANSWER: T

48. (T/F): A saddle joint is uniaxial.

ANSWER: F

49. (T/F): Movements of flexion and extension are permitted at synovial joints of the ellipsoid type.

ANSWER: T

50. (T/F): Abduction and adduction movements are permitted at synovial joints of the ellipsoid type.

ANSWER: T

51. (T/F): Inversion at the subtalar joint turns the sole of the foot laterally.

ANSWER: F

52. (T/F): Eversion at the subtalar joint turns the sole of the foot medially.

ANSWER: T

53. (T/F): Pronation of the forearm occurs when the palm of the hand is directed posteriorly.

ANSWER: T

54. (T/F): Supination of the forearm occurs when the palm of the hand is directed anteriorly.

ANSWER: T

55. (T/F): The movements of inversion and eversion occur at the talocrural joint.

ANSWER: F

56. (T/F):   Inversion and eversion occur at the subtalar and talocalcaneonavicular joints.

ANSWER: T

57. (T/F):   The movement called opposition is unique to the thumb.

ANSWER: T

58. (T/F):   Circumduction is permitted at spheroid joints.

ANSWER: T

59. (T/F):   Dorsiflexion is a movement unique to the subtalar joint.

ANSWER: F

60. (T/F):   Plantarflexion is a movement unique to the talocrural or ankle joint.

ANSWER: T

61. (T/F):   The Achilles tendon helps to limit the movement of plantarflexion.

ANSWER: F

62. (T/F):   Tension on muscles located in the anterior compartment of the leg help to limit the movement of plantarflexion.

ANSWER: T

63. (T/F):   Shapes of bones may limit movement at a joint.

ANSWER: T

64. (T/F):   Muscle tension and ligaments may limit motion at a joint.

ANSWER: T

65. (T/F): Apposition of soft tissue structures may limit motion at a joint.

ANSWER: T

INSTRUCTIONS: COMPLETE THE FOLLOWING:

66. Movement of a bone about its longitudinal axis is called:

ANSWER: rotation

67. The act of protruding or thrusting your mandible forward is termed:

ANSWER: protraction

68. The opposite of protraction is called:

ANSWER: retraction

69. A type of movement that is unique to the thumb is called:

ANSWER: opposition

70. The fixed point about which a lever is permitted to move is termed the:

ANSWER: fulcrum

71. The most common type of lever represented in the body is the:

ANSWER: third-class lever

72. Inflammatory involvement of a joint is known clinically as:

ANSWER: arthritis

73. A systemic disorder of uric acid metabolism that involves joints is known clinically as:

ANSWER: gout

74. A core of uric acid crystals surrounded by cells typical of an inflammatory response is called a:

ANSWER: tophus

75. An inflammed synovial membrane may be replaced by a mass of highly vascularized tissue containing lymphocytes and plasma cells called a:

ANSWER: pannus

76. In chronic rheumatoid arthritis, a joint often becomes totally immovable, or:

ANSWER: ankylosed

77. The acetabular notch is bridged by the:

ANSWER: transverse acetabular ligament

78. The ankle joint is also called the:

ANSWER: talocrural joint

79. The talocrural joint is strenghtened on its medial aspect by the:

ANSWER: deltoid ligament

80. More than 85% of the individuals with rheumatoid arthritis have an antibody present in their blood called:

ANSWER: rheumatoid factor (RF)

TEST BANK for CHAPTER 10: MUSCLE TISSUE

INSTRUCTIONS: ANSWER EACH QUESTION ACCORDING TO THE FOLLOWING KEY:

(A) Only 1 is correct
(B) Only 2 is correct
(C) Both are correct
(D) Neither are correct

1. The perimysium

    1. is continuous with the epimysium
    2. is continuous with the endomysium

ANSWER: C

2. The perimysium:

    1. surrounds bundles of muscle cells forming a fascicle
    2. is formed by connective tissue

ANSWER: C

3. The endomysium:

    1. is composed of connective tissue
    2. surrounds groups of muscle cells

ANSWER: A

4. The endomysium:

    1. surrounds individual muscle cells
    2. is continuous with the perimysium

ANSWER: C

5. Skeletal muscle cells:

    1. are multinucleated
    2. vary markedly in length

ANSWER: C

6. Myoblasts:

   1. arise from embryonic ectoderm
   2. are multinucleated

ANSWER: D

7. Myoblasts:

   1. arise from embryonic mesoderm
   2. are multinucelated

ANSWER: A

8. Myoblasts:

   1. arise from embryonic mesoderm
   2. possess a single nucleus

ANSWER: C

9. Early muscle stem cells:

   1. are called myoblasts
   2. arise from embryonic mesoderm

ANSWER: C

10. Satellite cells:

    1. are involved in muscle injury repair
    2. are undifferentiated precursor cells

ANSWER: C

11. Satellite cells:

    1. give rise to new muscle cells
    2. are typically located among all mature muscle cells

ANSWER: C

12. Myofilaments:

    1. actin
    2. myosin

13. The A band:

    1. overlapping thick and thin myofilaments
    2. contains the H band

ANSWER: C

14. The H band:

    1. lies within the A band
    2. formed by overlapping thin filaments

ANSWER: A

15. The H band:

    1. formed by overlapping thick and thin filaments
    2. lies within the I band

ANSWER: D

16. The H band:

    1. lies within the A band
    2. formed by overlapping thick myofilaments

ANSWER: C

17. The M band:

    1. lies within the H band
    2. connections between adjacent thin filaments

ANSWER: A

18. The M band:

    1. connections between adjacent thick filaments
    2. lies within the I band

ANSWER: A

19. The M band:

    1. lies within the H band
    2. connections between adjacent thick filaments

ANSWER: C

20. The I bands:

    1. located at the end of each sarcomere
    2. located adjacent to the Z lines

ANSWER: C

21. The I bands:

    1. located adjacent to each Z line
    2. contain thick myofilaments only

ANSWER: A

22. The I bands:

    1. contain thin myofilaments only
    2. appear darker than the A band

ANSWER: A

23. The I bands:

    1. appear lighter than the A band
    2. contain thick myofilaments only

ANSWER: A

24. The I bands:

    1. contain thin myofilaments only
    2. located adjacent to each Z line

ANSWER: C

25. Motor unit:

    1. group of muscle fibers that all receive their innervation from a single neuron
    2. constant in size

ANSWER: A

26. Myoneural junction:

    1. motor endplate
    2. neuromuscular junction

ANSWER: C

27. A microscopic space:

    1. neuromuscular junction
    2. myoneural cleft

ANSWER: C

28. Acetylcholine:

    1. released into myoneural cleft
    2. causes depolarization

ANSWER: C

29. Acetylcholinesterase:

    1. destroys acetylcholine
    2. prevents continuous depolarization

ANSWER: C

30. Myasthenia gravis (NG):

    1. autoimmune disorder
    2. affects mostly males

ANSWER: A

31. Myastheina gravis (MG):

    1. affects primarily females
    2. antibodies to acetylcholine receptors

ANSWER: C

32. Myasthenia gravis (MG):

    1. antibodies to acetylcholinesterase
    2. muscles become progesssively weaker

ANSWER: B

33. Myasthenia gravis (MG):

    1. initially involves face and neck muscles
    2. autoantibodies to acetylcholine receptors

ANSWER: C

TB.10:6

34. Simple muscle twitch:

    1. generated by a single depolarization
    2. lasts fraction of a second

ANSWER: C

35. Tetanic contraction:

    1. fusion of individual contractions
    2. longer than a simple muscle twitch

ANSWER: C

36. Slow twitch muscle cell:

    1. low fatigability
    2. short contraction time

ANSWER: A

37. Slow twitch muscle cell:

    1. short contraction time
    2. short relaxation time

ANSWER: D

38. Slow twitch muscle cell:

    1. long relaxation time
    2. low fat content

ANSWER: A

39. Slow twitch muscle cell:

    1. high capillary density
    2. high mitochondrial content

ANSWER: C

40. Fast twitch muscle cell:

    1. long contraction time
    2. long relaxation time

ANSWER: D

41. Fast twitch muscle cell:

    1. low mitochondrial content
    2. high capillary density

ANSWER: A

42. Fast twitch muscle cell:

    1. high myoglobin content
    2. low fat content

ANSWER: B

43. Fast twitch muscle cell:

    1. high fatigability
    2. short relaxation time

ANSWER: C

44. Myoglobin:

    1. a red pigment
    2. may damage kidneys if large quantities are liberated

ANSWER: C

45. Myoglobin:

    1. an intermediary in the transfer of oxygen from hemoglobin
    2. has a great oxygen-storing capacity

ANSWER: A

46. Myosin molecule:

    1. consists of eight polypeptide chains
    2. has a molecular weight of 40,000

ANSWER: D

47. Myosin filament:

    1. consists of about 200 myosin molecules
    2. composed partially of the protein, troponin

ANSWER: A

48. Actin filament:

    1. partly composed of troponin
    2. partly composed of tropomyosin

ANSWER: C

49. F-actin molecule:

    1. formed by two G-actin molecules
    2. a double stranded molecule

ANSWER: C

50. Rigor mortis:

    1. produces active contraction of muscle fibers
    2. produces stiffness

ANSWER: B

INSTRUCTIONS: TRUE - FALSE

51. (T/F): Muscle atrophy may result from lack of use of a muscle.

ANSWER: T

52. (T/F): Muscle power is defined as the amount of work achieved by a muscle over a specific time interval.

ANSWER: T

53. (T/F): Muscle power is influenced by rate of contraction.

ANSWER: T

54. (T/F): Muscle power is influenced by the velocity of contraction.

ANSWER: T

55. (T/F): One's nutritional status will influence one's muscular endurance.

ANSWER: T

56. (T/F): Recovery from aerobic energy production may require 24 to 48 hours.
ANSWER: T

57. (T/F): Smooth muscle cells are typically mononucleated.
ANSWER: T

58. (T/F): A nexus is a region of high electrical resistance.
ANSWER: F

59. (T/F): Smooth muscles cells have sarcomers which are very difficult to see with the light microscope.
ANSWER: F

60. (T/F): Smooth muscle cells may resopnd to norepinephrine.
ANSWER: T

61. (T/F): Smooth muscle is not capable of tonic contraction.
ANSWER: F

62. (T/F): Smooth muscle is capable of rhythmic contraction.
ANSWER: T

63. (T/F): Cytokinesis is dependent upon the presence of myosin.
ANSWER: T

64. (T/F): Smooth muscle typically contracts at a faster rate than other types of muscle.
ANSWER: F

65. (T/F): Calcium is important in the process of skeletal muscle contraction.
ANSWER: T

INSTRUCTIONS: COMPLETE THE FOLLOWING:

66. A condition representing the opposite of muscular hypertrophy is:

ANSWER: atrophy

67. Stiffening of the muscles upon death is called:

ANSWER: rigor mortis

68. That part of the actin filament which is capable of binding calcium is called:

ANSWER: troponin

69. An important feature of the myosin head is that in the presence calcium it has the ability to breakdown:

ANSWER: ATP

70. Resting muscle fibers derive most of their energy from the breakdown of:

ANSWER: fatty acids

71. The organelles in which beta oxidation occurs are called:

ANSWER: mitochondria

72. Fatty acids are broken down to acetate via a process called:

ANSWER: beta oxidation

73. During active work, muscles rely primarily upon _____ for their energy source.

ANSWER: glucose

74. Curare interfers with acetylcholine:

ANSWER: receptors

75. The charge differential between the interior and exterior of a resting cell sarcolemma is termed the:

ANSWER: resting potential

76. The term applied to the electrical changes responsible for transmitting information along the plasma membrane is:

ANSWER: depolarization

77. Myasthenia gravis (MG) is classified as an _____ disease.

ANSWER: autoimmune

78. A group of muscle fibers that all receive their innervation from a single neuron is called a:

ANSWER: motor unit

79. The enzyme responsible for the destruction of acetylcholine is:

ANSWER: acetylcholinesterase

80. The repeating functional unit of a skeletal muscle fiber is called a:

ANSWER: sarcomere

TEST BANK for CHAPTER 11: THE MUSCULAR SYSTEM

INSTRUCTIONS: ANSWER EACH QUESTION ACCORDING TO THE FOLLOWING KEY:

    (A) Only 1 is correct
    (B) Only 2 is correct
    (C) Both are correct
    (D) Neither are correct

1. Synergistic muscles:
    1. function to stabilize adjacent joints
    2. act in concert with prime movers

ANSWER: C

2. Muscles only:
    1. pull
    2. contract

ANSWER: C

3. Muscle shapes include:
    1. fusiform
    2. penniform

ANSWER: C

4. Muscle shapes include:
    1. bipenniform
    2. rhomboidal

ANSWER: C

5. A superficial muscle of the anterior abdominal wall:
    1. external abdominal oblique
    2. rectus abdominis

ANSWER: C

6. Muscle of the anterior thigh region:

    1. sartorius
    2. gracilis

ANSWER: A

7. Muscle of the posterior thigh region:

    1. biceps femoris
    2. quadriceps femoris

ANSWER: A

8. Muscle of the posterior leg:

    1. extensor digitorum brevis
    2. gastrocnemius

ANSWER: B

9. Muscle innervated by the facial nerve:

    1. temporalis
    2. mentalis

ANSWER: B

10. Muscle innervated by the facial nerve:

    1. buccinator
    2. posterior belly of digastric

ANSWER: C

11. Muscle innervated by the facial nerve:

    1. masseter
    2. anterior belly of the digastric

ANSWER: D

12. The platysma muscle:

    1. muscle of facial expression
    2. innervated by trigeminal nerve

ANSWER: A

13. Member of the erector spinae group of muscles:

    1. iliocostalis thoracis
    2. logissimus thoracis

ANSWER: C

14. Member of the subocciptal group of muscles:

    1. rectus capitis posterior minor
    2. obliquus capitis inferior

ANSWER: C

15. Functions in forced inspiration:

    1. rhomboideus major
    2. pectoralis major

ANSWER: C

16. Muscle that has its insertion upon the radius:

    1. biceps brachii
    2. coracobrachialis

ANSWER: A

17. Muscle insewrting upon the ulna:

    1. triceps brachii
    2. brachialis

ANSWER: B

18. Innervated by the axillary nerve:

    1. coracobrachialis
    2. triceps brachii

ANSWER: D

19. Innervated by the axillary nerve:

    1. deltoid
    2. brachialis

ANSWER: A

20. Innervated by the long thoracic nerve:

    1. serratus anterior
    2. serratus posterior inferior

ANSWER: A

21. Muscle inserting upon the humerus:

    1. teres minor
    2. subscapularis

ANSWER: C

22. Muscle inserting upon the humerus:

    1. teres major
    2. supraspinatus

ANSWER: C

23. Component of the rotator cuff:

    1. infraspinatus
    2. subscapularis

ANSWER: C

24. Component of the rotator cuff:

    1. infraspinatus
    2. deltoid

ANSWER: A

25. Innervated by the median nerve:

    1. palmaris longus
    2. flexor carpi radialis

ANSWER: C

26. Innervated by the radial nerve:

    1. extensor pollicis longus
    2. flexor pollicis longus

ANSWER: A

27. Pronates forearm and flexes forearm at elbow:

    1. pronator quadratus
    2. pronator teres

ANSWER: B

28. Muscle of the thenar eminence:

    1. opponens pollicis
    2. opponens digiti minimi

ANSWER: A

29. Flexor pollicis brevis:

    1. innervated by branch of median nerve
    2. muscle of thenar eminence

ANSWER: C

30. Flexor digiti minimi brevis:

    1. muscle of thenar eminence
    2. innervated by radial nerve

ANSWER: D

31. Has attachment to the iliac crest:

    1. latissimus dorsi
    2. internmal abdominal oblique

ANSWER: C

32. Has attachment to the iliac crest:

    1. qudratus lumborum
    2. rectus abdominis

ANSWER: A

33. Innervated by the superior gluteal nerve:

    1. gluteus medius
    2. gluteus maximus

ANSWER: A

34. Innervated by the inferior gluteal nerve:

    1. gluteus minimus
    2. gluteus maximus

ANSWER: B

35. Muscle that may have dual innervation:

    1. pectineus
    2. obturator externus

ANSWER: A

36. Gracilis muscle:

    1. innervated by femoral nerve
    2. abducts the thigh

ANSWER: D

37. Fibers originate from ischial tuberosity:

    1. semimembranosus
    2. semitendinosus

ANSWER: C

38. Innervated by tibial division of sciatic nerve:

    1. sartorius
    2. short head of biceps femoris

ANSWER: D

39. Innervated by deep branch of common peroneal nerve:

    1. tibialis anterior
    2. peroneus longus

ANSWER: A

40. Plantar flexes and everts the foot:

    1. tibialis anterior
    2. peroneus longus

ANSWER: B

41. Innervated by the medial plantar nerve:

    1. flexor digitorum brevis
    2. abductor hallucis

ANSWER: C

42. Plantar flexes the foot and flexes the knee:

    1. soleus
    2. gastrocnemius

ANSWER: B

43. Plantaris muscle:

    1. fibers originate on medial condyle of femur
    2. innervated by common peroneal nerve

Answer: D

44. Flexor digitorum accessorius muscle:

    1. some of its fibers originate from the long plantar ligament
    2. innervated by medial plantar nerve

ANSWER: C

45. Innervated by the lateral plantar nerve:

    1. flexor digiti minimi brevis
    2. adductor hallucis

ANSWER: C

46. Abducts the second through the fourth toes:

    1. dorsal interossei
    2. plantar interossei

ANSWER: A

47. Innervated by the pudendal nerve:

    1. bulbocavernosus
    2. sphincter urethrae

ANSWER: C

48. Under voluntary control:

    1. sphincter ani externus
    2. sphincter ani internus

ANSWER: A

49. Fibers originate from the mandible:

    1. styloglossus
    2. hyoglossus

ANSWER: D

50. Innervated by the hypoglossal nerve:

    1. styloglossus
    2. genioglossus

ANSWER: C

51. Fibers insert into fibrous raphe in posterior midline of the pahrynx:

    1. inferior constrictor
    2. superior constrictor

ANSWER: C

52. Innervated by the pharyngeal plexus:

    1. palatopharyngeus
    2. stylopharyngeaus

ANSWER: A

53. Innervated by the external laryngeal branch of the vagus nerve:

    1. cricothyroid
    2. lateral cricoarytenoid

ANSWER: A

54. Fibers originate on the eustachian tube:

    1. palatopharyngeus
    2. salpingopharyngeus

ANSWER: B

55. Depresses the hyoid bone:

    1. sternohyoid
    2. omohyoid

ANSWER: C

56. Tendon forms part of the medial hamstring tendons:

    1. semimembranosus
    2. semitendinosus

ANSWER: C

57. Innervated by the tibial division of the sciatic nerve:

    1. semitendinosus
    2. long head of biceps femoris

ANSWER: C

58. Lateral hamstring tendon:

    1. long head of biceps femoris muscle
    2. short head of biceps femoris muscle

ANSWER: C

59. Flexes arm at shoulder:

    1. coracobrachialis
    2. triceps brachii

ANSWER: A

60. Innervated by radial nerve:

    1. brachioradialis
    2. deltoid

ANSWER: A

PROVIDE THE CORRECT INNERVATION FOR THE FOLLOWING MUSCLES
ACCORDING TO THE KEY PROVIDED BELOW:

        (A) - median nerve
        (B) - radial nerve
        (C) - axillary nerve
        (D) - musculocutaneous nerve
        (E) - none of the above nerves

61. Coracobrachialis

ANSWER: D

62. Triceps brachii

ANSWER: B

63. Brachioradialis

ANSWER: B

64. Biceps brachii

ANSWER: D

65. Deltoid

ANSWER: C

66. Brachialis

ANSWER: D

67. Teres major

ANSWER: E

68. Pectoralis minor

ANSWER: E

69. Supraspinatus

ANSWER: E

70. Infraspinatus

ANSWER: E

PROVIDE THE CORRECT INNERVATION FOR THE FOLLOWING MUSCLES ACCORDING TO THE KEY PROVIDED BELOW:

    (A) - median nerve
    (B) - radial nerve
    (C) - ulnar nerve

71. Extensor indicis

ANSWER: B

72. Extensor pollicis longus

ANSWER: B

73. Extensor digiti minimi

ANSWER: B

74. Pronator quadratus

ANSWER: A

75. Abductor pollicis longus

ANSWER: B

76. Flexor pollicis longus

ANSWER: A

77. Flexor carpi ulnaris

ANSWER: C

78. Palmaris longus

ANSWER: A

79. Palmaris brevis

ANSWER: C

80. Opponens digiti minimi

ANSWER: C

81. Opponens pollicis

ANSWER: A

82. Abductor digiti minimi

ANSWER: C

83. Abductor pollicis brevis

ANSWER: A

84. Fourth and fifth lumbricals

ANSWER: C

85. Second and third lumbricals

ANSWER: A

86. Palmar interossei

ANSWER: C

87. Dorsal interossei

ANSWER: C

PROVIDE THE CORRECT INNERVATION FOR THE FOLLOWING MUSCLES
ACCORDING TO THE KEY PROVIDED BELOW:

  (A) — femoral nerve
  (B) — obturator nerve
  (C) — inferior gluteal nerve
  (D) — superior gluteal nerve

88. Tensor fasciae latae

ANSWER: D

89. Adductor longus

ANSWER: B

90. Adductor brevis

ANSWER: B

91. Gluteus maximus

ANSWER: C

92. Gluteus medius

ANSWER: D

93. Gluteus minimus

ANSWER: D

94. Iliacus

ANSWER: A

95. Obturator externus

ANSWER: B

96. Pectineus (when not bilaminar)

ANSWER: A

97. Vastus lateralis

ANSWER: A

98. Vastus medialis

ANSWER: A

99. Vastus intermedius

ANSWER: A

100. Quadriceps femoris

ANSWER: A

PROVIDE THE CORRECT INNERVATION FOR THE FOLLOWING MUSCLES
ACCORDING TO THE KEY PROVIDED BELOW:

(A) - deep branch of common peroneal nerve
(B) - superficial branch of common peroneal nerve
(C) - tibial nerve
(D) - medial plantar nerve
(E) - lateral plantar nerve

101. Flexor hallucis brevis

ANSWER: D

102. Flexor digiti minimi brevis

ANSWER: E

103. Plantar interossei

ANSWER: E

104. Adductor hallucis

ANSWER: E

105. Dorsal interossei

ANSWER: E

106. Flexor digitorum brevis

ANSWER: D

107. Gastrocnemius

ANSWER: C

108. Peroneus tertius

ANSWER: A

109. Plantaris

ANSWER: C

110. Soleus

ANSWER: C

111. Peroneus brevis

ANSWER: B

112. Peroneus longus

ANSWER: B

113. Tibialis anterior

ANSWER: A

114. Flexor hallucis longus

ANSWER: C

115. Extensor hallucis longus

ANSWER: A

116. Extensor digitorum longus

ANSWER: A

INSTRUCTIONS: TRUE - FALSE

117. (T/F): The abductor digiti minimi is a stronger flexor than an abductor.

ANSWER: T

118. (T/F): The lumbricals of the foot arise from the four tendons of the flexor digitorum brevis muscle.

ANSWER: F

119. (T/F): The soleus muscle dorsiflexes the foot.

ANSWER: F

120. (T/F): The peroneus tertius is frequently described as being part of the extensor digitorum longus muscle.

ANSWER: T

121. (T/F): The pectineus muscle may receive innervation from the obturator nerve if the muscle is bilaminar.

ANSWER: T

122. (T/F): The pectineus is an abductor of the thigh.

ANSWER: F

123. (T/F): The pectineus muscle originates from the pubis.

ANSWER: T

124. (T/F): The gluteus maximus is considered to be the primary extensor of the thigh.

ANSWER: T

125. (T/F): Quadratus femoris medially rotates the thigh.

ANSWER: F

126. (T/F): Together with the iliacus muscle, the psoas major is often referred to as the iliopsoas muscle.

ANSWER: T

127. (T/F): The piriformis muscle laterally rotates the extended thigh.

ANSWER: T

128. (T/F): The tensor fasciae latae may extend and laterally rotate the leg at the knee.

ANSWER: T

129. (T/F): Fibers of the flexor carpi ulnaris originate from the humerus and the ulna.

ANSWER: T

130. (T/F): Flexor digitorum profundus originates in part from the interosseous membrane.

ANSWER: T

TEST BANK for CHAPTER 12: NERVOUS SYSTEM: BASCI ORGANIZATION
AND FUNCTION

INSTRUCTIONS: ANSWER EACH QUESTION ACCORDING TO THE FOLLOWING KEY:

    (A) Only 1 is correct
    (B) Only 2 is correct
    (C) Both are correct
    (D) Neither are correct

1. A division of the central nervous system:

    1. brain
    2. spinal cord

ANSWER: C

2. A division of the nervous system:

    1. central nervous system
    2. peripheral nervous system

ANSWER: C

3. Division of the peripheral nervous system:

    1. somatic portion
    2. autonomic portion

ANSWER: C

4. Somatic division of the PNS:

    1. has afferent nerves
    2. has efferent nerves

ANSWER: C

5. Efferent neurons:

    1. only in the autonomic division of the PNS
    2. only in the somatic division of the PNS

ANSWER: D

6. Glial cells:

    1. provide protection
    2. provide support

ANSWER: C

7. Glial cells:

    1. provide support
    2. give rise to 40 to 45% of all intracranial tumors

ANSWER: C

8. Glial cells:

    1. secrete a tough matrix
    2. collectively referred to as the neuroglia

ANSWER: B

9. Glial cells:

    1. more numerous than neurons
    2. do not secret a tough matrix

ANSWER: C

10. A glial cell type within the CNS:

    1. ependymal cell
    2. Schwann cell

ANSWER: A

11. Ependymal cells:

    1. glial cells of the CNS
    2. line cavities of brain

ANSWER: C

12. Microglia:

    1. often called 'brain macrophages'
    2. active in response to inflammation

ANSWER: C

13. Schwann cells:

    1. glial cells of the CNS
    2. form myelin sheaths

ANSWER: B

14. Cells forming myelin sheaths:

    1. Schwann cells
    2. Oligodendrocytes

ANSWER: C

15. The neuron:

    1. structural/functional unit of the nervous system
    2. long cytoplasmic processes

ANSWER: C

16. Neuron cell body:

    1. perikaryon
    2. soma

ANSWER: C

17. Nissil substance consists of:

    1. smooth endoplasmic reticulum
    2. Golgi apparatus

ANSWER: D

18. Nissil substance:

    1. located in neuron cell body
    2. consists partly of free ribosomes

ANSWER: C

19. Nissil substance:

    1. functions in protein synthesis
    2. consists in part of rough endoplasmic reticulum

ANSWER: C

TB.12:4

20. Microfilaments of neurons:

    1. anchored to plasm membrane by the protein called fodrin
    2. are composed of actin

ANSWER: C

21. Components of the neuron cytoskeleton:

    1. microtubules
    2. microfilaments

ANSWER: C

22. Function of the neuron cytoskeleton:

    1. maintains cell shape
    2. transports materials/organelles

ANSWER: C

23. Microfilaments in neurons:

    1. are composed of myosin
    2. do not dissociate

ANSWER: D

24. Dendrites:

    1. are specialized to transmit neural impulses
    2. represent extensions of the cytoplasm

ANSWER: B

25. Dendrites:

    1. conduct neural impulses toward the cell body
    2. are dotted with dendritic spines

ANSWER: C

26. The cytoplasm of the axon is:

    1. enveloped by the axolemma
    2. called the axoplasm

ANSWER: C

27. Axons may:

    1. be more than three feet in length
    2. arise from the axon hillock

ANSWER: C

28. Axons:

    1. may give off collaterals
    2. give rise to axon terminals

ANSWER: C

29. Transports materials away from the cell body:

    1. axonal transport
    2. axoplasmic flow

ANSWER: C

30. Movement of materials that requires ATP and calcium:

    1. axonal transport
    2. axoplasmic flow

ANSWER: A

31. Produced by Schwann cells:

    1. myelin sheath
    2. neurilemma

ANSWER: C

32. Transport of materials toward the cell body:

    1. axonal transport
    2. axoplasmic flow

ANSWER: A

33. Transports materials at a rate of approximately 400 mm per day:

    1. axoplasmic flow
    2. axonal transport

ANSWER: B

TB.12:6

34. Myelinated axons:

    1. generally larger than 2 micrometers in diameter
    2. exhibit nodes of Ranvier

ANSWER: C

35. Myelin:

    1. responsible for the white color of white matter
    2. formed exclusively by Schwann cells

ANSWER: A

36. Multiple sclerosis:

    1. disease characterized by myelin deterioration
    2. myelin relpaced by astrocytes

ANSWER: C

37. Disease in which myelin deteriorates or is destroyed:

    1. multiple sclerosis
    2. Guillain-Barre syndrome

ANSWER: C

38. Agent affecting myelin or myelinization process:

    1. hexachloraphene
    2. organic tin compounds

ANSWER: C

39. Cellular sheaths (neurilemmas):

    1. found only in the CNS
    2. serve as guides for regenerating axons

ANSWER: B

40. Bipolar neurons have only one:

    1. axon
    2. dendrite

ANSWER: C

TB.12:7

41. Bipolar neurons are found in the:

   1. retina
   2. inner ear

ANSWER: C

42. Unipolar neurons:

   1. have just one cell process
   2. are typically sensory neurons

ANSWER: C

43. Neurons linking sensory and motor neurons:

   1. interneurons
   2. association neurons

ANSWER: C

44. The endoneurium:

   1. surrounds a group of several axons
   2. is composed of connective tissue

ANSWER: B

45. The perineurium:

   1. envelops individual axons
   2. envelops groups of axons and forms fasicles

ANSWER: D

46. The perneurium:

   1. is formed from connective tissue
   2. surrounds the nerve and holds the fasicles together

ANSWER: A

47. Collections of cell bodies within the CNS:

   1. centers
   2. nuclei

ANSWER: C

TB.12:8

48. Collections of cell bodies outside of the CNS:

   1. ganglia
   2. nuclei

ANSWER: A

49. Every neural response depends on:

   1. integration
   2. reception

ANSWER: C

50. The difference in potential between two sides of a membrane:

   1. resting potential
   2. membrane potential

ANSWER: C

51. The resting potential of a neuron:

   1. is measured in millivolts
   2. amounts to approximately 200 millivolts

ANSWER: A

52. The resting potential of a neuron:

   1. requires energy be utilized
   2. is due in part to the presence of large protein anions within the cell

ANSWER: C

53. May alter the resting potential of a neuron:

   1. chemical stimulus
   2. electrical stimulus

ANSWER: C

54. Voltage-activated ion channels:

   1. are sensitive to transmembrane voltage
   2. open when the threshold level is reached

ANSWER: C

55. Neurons:

   1. are highly excitable cells
   2. convert stimuli into neural impulses

ANSWER: C

INSTRUCTIONS: COMPLETE THE FOLLOWING:

56. The sharp rise and fall of the action potential are collectively referred to as the:

ANSWER: spike

57. A neural impulse is transmitted as a wave of:

ANSWER: depolarization

58. By the time the action potential has moved a few millimeters, the membrane over which it just passed begins to:

ANSWER: repolarize

59. An axon membrane cannot transmit another action potential during the interval known as the absolute:

ANSWER: refractory period

60. The smooth, progressive impulse transmission characteristic of unmyelinated neurons is called:

ANSWER: continuous conduction

INSTRUCTIONS: TRUE - FALSE

61. (T/F): A neuron obeys an all-or-none-law.

ANSWER: T

62. (T/F): A neuron that begins at a synapse is called the postsynaptic neuron.

ANSWER: T

63. (T/F): An impulse may not be electrically transmitted from the presynaptic neuron to the postsynaptic neuron.

ANSWER: F

64. (T/F): Electrical and chemical synapses are known to exist.

ANSWER: T

65. (T/F): Most synapses in the body are thought ot be electrical synapses.

ANSWER: F

66. (T/F): The space separating pre- and postsynaptic cells is called the synaptic cleft.

ANSWER: T

67. (T/F): Neurotransmitter substance is stored within synaptic vesicles of the postsynaptic neuron.

ANSWER: F

68. (T/F): Calcium induces synaptic vesicles to undergo fusion.

ANSWER: T

69. (T/F): Not all neurotransmitters are inactivated by enzymes.

ANSWER: T

70. (T/F): Some neurotransmitter-receptor combinations result in hyperpolarization of the postsynaptic membrane.

ANSWER: T

71. (T/F): EPSPs may be added together, a process called summation.

ANSWER: T

72. (T/F): A neuron that is facilitated is sufficiently depolarized to conduct an impulse.

ANSWER: F

73. (T/F): A single neuron secretes only one type of neurotransmitter.

ANSWER: F

74. (T/F): Cells releasing acetylcholine are designated as cholinergic neurons.

ANSWER: T

75. (T/F): Type B fibers are the smallest and slowest conducting nerve fibers.

ANSWER: F

76. (T/F): Convergence means that a single neuron is controlled by signals arriving from tow or more presynaptic neurons.

ANSWER: T

77. (T/F): The arrangement in which a single presynaptic neuron stimulates many postsynaptic cells is called divergence.

ANSWER: T

78. (T/F): After discharge may prolong a response.

ANSWER: T

79. (T/F): A reverberating circuit is an example of positive feedback.

ANSWER: T

80. (T/F): Neurons do not discharge spontaneously.

ANSWER: F

TEST BANK for CHAPTER 13: THE CENTRAL NERVOUS SYSTEM

INSTRUCTIONS: ANSWER EACH QUESTION ACCORDING TO THE FOLLOWING KEY:

(A) Only 1 is correct
(B) Only 2 is correct
(C) Both are correct
(D) Neither are correct

1. The subdural space:

    1. is a potential space
    2. separates the dura from the arachnoid

ANSWER: C

2. Regarding the pia mater:

    1. is an avascular membrane
    2. adheres closely to the inner aspect of the skull

ANSWER: D

3. The subarachnoid space:

    1. contains cerebrospinal fluid (CSF)
    2. lies between the arachnoid and the dura

ANSWER: A

4. The tentorium cerebelli:

    1. represents a partition formed from dura
    2. is attached to the crista galli

ANSWER: A

5. The falx cerebri:

    1. lies between the cerebellum and the cerebral hemispheres
    2. is formed largely from pia and arachnoid

ANSWER: D

6. The central nervous system (CNS):

    1. consists of the brain and spinal cord
    2. is covered by the meninges

ANSWER: C

7. The choroid plexuses:

    1. produce CSF
    2. project into the ventricles

ANSWER: C

8. The CSF contains:

    1. glucose
    2. urea

ANSWER: C

9. The CSF contains:

    1. water
    2. red blood cells

ANSWER: A

10. The arachnoid granulations:

    1. reabsorb CSF into the blood
    2. are located only in the lateral ventricles

ANSWER: A

11. The cerebrospinal fluid (CSF):

    1. functions in nutrition of the CNS
    2. cushions the CNS and protects it from traumatic injury

ANSWER: C

12. During embryonic development the prosencephalon:

    1. gives rise to the telecephalon and diencephalon
    2. is known as the hindbrain

ANSWER: A

13. During embryonic development the rhombencephalon divides to form the:

    1. diencephalon
    2. myelencephalon

ANSWER: B

14. The metencephalon ultimately gives rise to the:

    1. cerebrum
    2. medulla

ANSWER: D

15. The diencephalon ultimately gives rise to the:

    1. thalamus
    2. cerebellum

ANSWER: A

16. A primary brain vesicle during embryonic development:

    1. mesencephalon
    2. rhombencephalon

ANSWER: C

17. The fourth ventricle communicates with the:

    1. central canal of the spinal cord
    2. cerebral aqueduct

ANSWER: C

18. The medulla oblongata:

    1. is the most inferior portion of the brainstem
    2. contains a cavity called the third ventricle

ANSWER: A

19. Regarding the medulla oblongata:

    1. contains gray matter
    2. contains white matter

ANSWER: C

20. The pryamids:

    1. are located on the anterior surface of the medulla
    2. consist primarily of white matter

ANSWER: C

21. The pryamidal tracts:

    1. contain anterior corticospinal fibers
    2. represent major voluntary motor pathways

ANSWER: C

23. The decussation of the pyramids:

    1. occurs just superior to the junction of the medulla with the spinal cord
    2. includes fibers of the lateral corticospinal tracts

ANSWER: C

24. Located within the medulla oblongata:

    1. nucleus cuneatus
    2. nucleus gracilis

ANSWER: C

25. Contains a portion of the reticular formation:

    1. medulla oblongata
    2. spinal cord

ANSWER: C

26. Nucleus located within the medulla oblongata:

    1. 10th cranial nerve (CN X)
    2. 8th cranial nerve (CN VIII)

ANSWER: A

27. Regarding the pons:

    1. located on the anterior (ventral) surface of the brainstem
    2. contains centers that help regulate respiration

ANSWER: C

28. Regarding the midbrain:

    1. contains a cavity called the third ventricle
    2. shortest portion of the brainstem

ANSWER: B

29. A component of the cerebral peduncle:

    1. substantia nigra
    2. nucleus cuneatus

ANSWER: A

30. A component of the cerebral peduncle:

    1. crus cerebri
    2. cerebral aqueduct

ANSWER: A

31. The tegmental portion of the cerebral peduncle contains:

    1. part of the reticular formation
    2. the nucleus of the 8th cranial nerve (CN VIII)

ANSWER: A

32. Located within the tegmental portion of the cerebral peduncle:

    1. nucleus of 3rd cranial nerve (CN III)
    2. nucleus of 4th cranial nerve (CN IV)

ANSWER: C

33. The red nucleus:

    1. lies within the reticular formation
    2. is located within the midbrain

ANSWER: C

34. The red nucleus contains neurons that:

    1. receive information from the cerebellum
    2. send information to the spinal cord
       via the rubrospinal tract

ANSWER: C

35. The roof of the midbrain:

    1. lies anterior to the cerebral aqueduct
    2. is called the tectum

ANSWER: B

36. The corpora quadrigemina:

    1. are located in the tectum
    2. form a portion of the midbrain

ANSWER: C

37. Regarding the inferior colliculi:

    1. represent relay centers for visual information
    2. are located within the pons

ANSWER: D

38. Regarding the superior colliculi:

    1. located in the tectum
    2. form part of the corpora quadrigemina

ANSWER: C

39. Regarding the diencephalon:

    1. its cavity is called the third ventricle
    2. contains the thalamus

ANSWER: C

40. The thalamus:

   1. is a major relay center
   2. only receives information from descending fiber tracts

ANSWER: A

41. Thalamic nuclei relaying motor information:

   1. ventral lateral nuclei
   2. ventral anterior nuclei

ANSWER: C

42. Thalamic nuclei relaying visual information:

   1. lateral geniculate nuclei
   2. medial geniculate nuclei

ANSWER: A

43. Regarding the hypothalamus:

   1. contains the supraoptic nuclei
   2. contains the paraventricular nuclei

ANSWER: C

44. Contains cells which produce the hormone, oxytocin:

   1. hypothalamus
   2. suproptic nuclei

ANSWER: A

45. Regarding the hypothalamus:

   1. contains a thirst regulating center
   2. contains a temperature regulating center

ANSWER: C

46. The hypothalamus is:

   1. involved with temperature regulation
   2. located in the floor of the fourth ventricle

ANSWER: A

47. The hypothalamus influences:

    1. sexual behavior
    2. the fucntion of the autonomic nervous system

ANSWER: C

48. The two hemispheres of the cerebellum are:

    1. connected by the vermis
    2. partially separated by the falx cerebelli

ANSWER: C

49. Regarding the cerebellum:

    1. its major output pathway is through the inferior cerebellar peduncle
    2. sends information to the red nucleus of the midbrain

ANSWET: B

50. Functions of the cerebellum:

    1. plays a role in maintaining posture
    2. helps maintain muscle tone

ANSWER: C

INSTRUCTIONS: TRUE - FALSE

51. (T/F): An individual with damage to the cerebellum may develop a condition called ataxia.

ANSWER: T

52. (T/F): The cerebral cortex is a thin layer and is composed primarily of white matter.

ANSWER: F

53. (T/F): Function of the flocculonodular lobe of the cerebellum is concerned with equlibrium and posture.

ANSWER: T

54. (T/F): Shallow grooves of the cerebral hemispheres are referred to as gyri.

ANSWER: F

55. (T/F): Neocortex may be distinguished histologically from paleocortex.

ANSWER: T

56. (T/F): Neocortex in humans is composed of six distinct layers.

ANSWER: T

57. (T/F): Myelinated fibers of the cerebrum may be classified as associative, commissural, or projection types.

ANSWER: T

58. (T/F): The caudate and lentiform nuclei contribute to the formation of the baszl ganglia.

ANSWER: T

59. (T/F): Each caudate nucleus consists of the putamen and globus pallidus.

ANSWER: F

60. (T/F): The caudate and lentiform nuclei are referred to collectively as the corpus striatum.

ANSWER: F

61. (T/F): According to Brodmann's classification scheme, areas 4, 6 and 8 are designated as primary sensory regions of the cerebral cortex.

ANSWER: T

62. (T/F): The prefrontal area of the cerebral cortex is also known as an association area.

ANSWER: T

63. (T/F): The precentral gyrus contains Brodmann's sensory areas 1, 2 and 3.

ANSWER: T

64. (T/F): The visual cortex is located in the anterior portion of the temporal lobe.

ANSWER: F

65. (T/F): The limbic lobe is thought to function as a link between cognitive and emotional mechanisms.

ANSWER: T

66. (T/F): The clinical term for an impairment in the reception, manipulation, or expression of words is called aphasia.

ANSWER: T

67. (T/F): The limbic system is composed of structures in the cerebrum as well as in the diencephalon.

ANSWER: T

68. (T/F): The amygdala represents a compllex of nuclei within the cerebrum.

ANSWER: T

69. (T/F): Delta waves are slow, large waves associated with stressful situations.

ANSWER: F

70. (T/F): The reticular activating system is also known as the arousal system.

ANSWER: T

INSTRUCTIONS: COMPLETE THE FOLLOWING:

71. The caudal end of the spinal cord narrows to a sharp tip called the:

ANSWER: conus medullaris

72. An extension of the pia mater from the conus medullaris to the tip of the coccyx is called the:

ANSWER: filum terminale

73. In each half of the spinal cord the white matter is arranged into three columns called:

ANSWER: funiculi

74. The spinothalamic tracts convey information relating to pain and:

ANSWER: temperature

75. The lateral and anterior corticospinal tracts are also known as the:

ANSWER: pyramidal tracts

76. A reflex arc which does not require the participation of the brain is called a:

ANSWER: spinal reflex

77. A stretch reflex is also designated as a:

ANSWER: monosynaptic reflex

78. The stimulation of one muscle and the simultaneous inhibition of antagonistic muscles is called:

ANSWER: reciprocal inhibition

79. Withdrawl reflexes are designated as:

ANSWER: polysynaptic reflexes

80. A simple neural pathway linking receptor, CNS, and effector is called a:

ANSWER: reflex arc

TEST BANK for CHAPTER 14: THE PERIPHERAL NERVOUS SYSTEM: SOMATIC SYSTEM

INSTRUCTIONS: COMPLETE THE FOLLOWING:

1. Cell bodies of most sensory neurons of the fifth cranial nerve are located in the:

ANSWER: trigeminal ganglion

2. Bell's palsy is a common disorder of the:

ANSWER: 7th cranial nerve (Facial nerve)

3. Tic douloureux is also known by the name of:

ANSWER: trigeminal neuralgia

4. The ventral rami of several spinal nerves may form networks called:

ANSWER: plexuses

5. The area of skin supplied by any one spinal nerve through both its rami is called a segment, or:

ANSWER: dermatome

6. Acute inflammation of the nervous system caused by the herpes zoster virus is termed:

ANSWER: shingles

7. The number of pairs of cervical spinal nerves is:

ANSWER: eight

8. The number of pairs of thoracic spinal nerves is:

ANSWER: twelve

9. The number of pairs of lumbar spinal nerves is:

ANSWER: five

10. The number of pairs of sacral spinal nerves is:

ANSWER: five

11. Immediately before the dorsal root unites with the spinal cord it is marked by a swelling called the:

ANSWER: dorsal root ganglion

12. Dorsal and ventral roots unite at the intervertebral foramen to form the:

ANSWER: spinal nerve

13. The trigeminal nerve (CN. V) gives rise to three main branches, the ophthalmic division, maxillary division, and:

ANSWER: mandibular division

14. Deep branches of the cervical plexus supplying motor function to the diaphragm are the:

ANSWER: phrenic nerves

15. The transverse cervical nerves arise from the _____ plexus.

ANSWER: cervical

16. The Phrenic nerves arise from the _____ plexus.

ANSWER: cervical

17. The greater auricular nerves arise from the _____ plexus.

ANSWER: cervical

18. The thoracodorsal nerve emerges from the _____ cord of the brachial plexus.

ANSWER: posterior

19. The ulnar nerve emerges from the _____ cord of the brachial plexus.

ANSWER: medial

20. The lateral pectoral nerve emerges from the _____ cord of the brachial plexus.

ANSWER: lateral

21. The musculocutaneous nerve emerges from the _____ cord of the brachial plexus.

ANSWER: lateral

22. The radial nerve emerges from the _____ cord of the brachial plexus.

ANSWER: posterior

23. The lower subscapular nerve emerges from the _____ cord of the brachial plexus.

ANSWER: posterior

24. The first eleven thoracic spinal nerves are also called the:

ANSWER: intercostal nerves

25. The femoral and obturator nerves arise from the _____ plexus.

ANSWER: lumbar

26. The genitofemoral nerves arise from the _____ plexus.

ANSWER: lumbar

27. The lesser occipital nerves arise from the _____ plexus.

ANSWER: cervical

28. The sciatic nerve arises from the _____ plexus.

ANSWER: sacral

29. The superior gluteal nerve arises from the _____ plexus.

ANSWER: sacral

30. The sciatic nerve consists of a common peroneal and a _____ division.

ANSWER: tibial

31. The sciatic nerve is composed of fibers from the ventral primary rami of spinal nerves L-_____ through S-_____.

ANSWER: L-4 through S-3

32. The femoral nerve is composed of fibers from the ventral primary rami of spinal nerves L-_____ through L-_____.

ANSWER: L-2 through L-4

33. The pudendal nerve is derived from the _____ plexus.

ANSWER: sacral

34. The dorsal root of a spinal nerve is composed of _____ fibers.

ANSWER: sensory (afferent)

35. The _____ nerve is the largest nerve emerging from the lumbar plexus.

ANSWER: femoral

36. There are _____ pairs of cranial nerves.

ANSWER: twelve

37. The _____ nerve is responsible for innervating the muscles of facial expression.

ANSWER: facial (7th cranial)

38. The _____ nerve innervates the superior oblique muscle of the orbit.

ANSWER: trochlear (4th cranial)

39. The _____ nerve innervates the lateral rectus muscle of the orbit.

ANSWER: abducens (6th cranial)

40. The superior rectus muscle of the orbit is innervated by the _____ nerve.

ANSWER: oculomotor (3rd cranial)

41. The inferior oblique and medial rectus muscles are innervated by the _____ nerve.

ANSWER: oculomotor (3rd cranial)

42. The optic nerves contain only _____ fibers.

ANSWER: sensory

43. Neurons pass caudally from the optic chiasma forming the optic:

ANSWER: tracts

44. The only cranial nerves to emerge from the dorsal surface of the brainstem are the:

ANSWER: trochlear nerves (CN IV)

45. Within the _____ _____, neurons from the medial portion of each retina cross to the opposite side; those from the lateral portion remain on the same side.

ANSWER: optic chiasma

INSTRUCTIONS: TRUE - FALSE

46. (T/F): Axons that innervate smooth muscle cells in the eyeball pass through the ciliary ganglion.

ANSWER: T

47. (T/F): The trochlear nerves are both motor and sensory.

ANSWER: T

48. (T/F): The trigeminal nerves are mixed nerves.

ANSWER: T

49. (T/F): The olfactory system is the only sensory system that passes directly to the cerebral cortex without synapsing in the thalamus.

ANSWER: T

50. (T/F): The oculomotor nerves are mixed nerves.

ANSWER: T

51. (T/F): Tic douloureux may cause excruciating pain over the distribution of the abducens nerve.

ANSWER: F

52. (T/F): The facial nerve conveys information regarding taste from the posterior one-third of the tongue.

ANSWER: F

53. (T/F): Bell's palsy is usually accompanied by a complete spontaneous recovery within several months.

ANSWER: T

54. (T/F): The vagus nerves innervate structures in the thorax and abdomen.

ANSWER: T

55. (T/F): The vestibular branch of the eighth cranial nerve conveys information regarding equilibrium and orientation of the head in space.

ANSWER: T

56. (T/F): The vagus nerves exit the skull through the jugular foramina.

ANSWER: T

57. (T/F): Motor fibers of the facial nerve supply the lacrimal, submaxillary, and sublingual glands.

ANSWER: T

58. (T/F): The hypoglossal nerves exit the skull through the jugular foramina.

ANSWER: F

59. (T/F): The hypoglossal nerves innervate the muscles of the tongue.

ANSWER: T

60. (T/F): The dorsal root ganglion contains cell bodies of motor neurons.

ANSWER: F

61. (T/F): The ventral root of a spinal nerve contains motor (efferent) fibers.

ANSWER: T

62. (T/F): A spinal nerve gives rise to dorsal and ventral rami.

ANSWER: T

63. (T/F): The ventral rami of spinal nerves T2 to T11 form the brachial plexus.

ANSWER: F

64. (T/F): Branches from the cervical plexus communicate with the seventh and eighth cranial nerves.

ANSWER: F

65. (T/F): The median, ulnar, and radial nerves all arise from the brachial plexus.

ANSWER: T

66. (T/F): The brachial plexus may receive some fibers from C4 and T2.

ANSWER: T

67. (T/F): The brachial plexus consists, in part, of three trunks.

ANSWER: T

68. (T/F): The brachial plexus consists, in part, of four cords.

ANSWER: F

69. (T/F): Ventral rami of spinal nerves C5 and C^ join to form the upper trunk of the brachial plexus.

ANSWER: T

70. (T/F): The middle trunk of the brachial plexus consists of the ventral ramus of spinal nerve C7.

ANSWER: T

72. (T/F): The femoral nerve innervates the skin and anterior muscles of the thigh.

ANSWER: T

73. (T/F): The lumbar plexus consists of dorsal and ventral rami of spinal nerves.

ANSWER: F

74. (T/F): The largest nerve in the body is the femoral nerve.

ANSWER: F

75. (T/F): Sciatica may be related to degenerative joint disease in the lumbosacral region.

ANSWER: T

76. (T/F): The phrenic nerves receive fibers from spinal nerves C3, C4, and C5.

ANSWER: T

77. (T/F): The ilioinguinal and iliohypogastric nerves are derived from the lumbar plexus of nerves.

ANSWER: T

78. (T/F): The latissimus dorsi muscle receives its innervation from the thoracodorsal nerve.

ANSWER: T

79. (T/F): The upper subscapular nerve arises from the lateral cord of the brachial plexus.

ANSWER: F

80. (T/F): The deltopid and teres minor muscles are innervated by the radial nerve.

ANSWER: F

INSTRUCTIONS: ANSWER EACH QUESTION ACCORDING TO THE FOLLOWING KEY:

(A) Only 1 is correct
(B) Only 2 is correct
(C) Both are correct
(D) Neither are correct

81. Regarding the radial nerve:

1. arises from the brachial plexus
2. innervates the latissimus dorsi muscle

ANSWER: A

82. Regarding the musculocutaneous nerve:

1. arises from the medial cord of the brachial plexus
2. innervates the triceps brachii muscle

ANSWER: A

83. Regarding the long thoracic nerve:

1. arises from the medial cord of the brachial plexus
2. innervates the subclavius muscle

ANSWER: D

84. Regarding the dorsal scapular nerve:

    1. contains fibers primarily from C2 and C3
    2. innervates the serratus anterior muscle

ANSWER: D

85. Regarding the axillary nerve:

    1. arises from the lateral cord of the brachial plexus
    2. innervates the deltoid muscle

ANSWER: B

86. Regarding the posterior cord of the brachial plexus:

    1. gives rise to the axillary nerve
    2. gives rise to the radial nerve

ANSWER: C

87. Regarding the medial cord of the brachial plexus:

    1. gives rise to the musculocutaneous nerve
    2. gives rise to the long thoracic nerve

ANSWER: D

88. Regarding the posterior cord of the brachial plexus:

    1. gives rise to the thoracodorsal nerve
    2. gives rise to the upper subscapular nerve

ANSWER: C

89. Regarding the obturator nerve:

    1. arises from the sacral plexus
    2. is sensory to the skin on the medial
       aspect of the thigh

ANSWER: B

90. Regarding the lumbar plexus:

    1. gives rise to the lateral femoral cutaneous nerve
    2. gives rise to the genitofemoral nerve

ANSWER: C

TEST BANK for CHAPTER 15: PERCEPTION OF SENSATION, MOTOR CONTROL, AND INFORMATION PROCESSING

INSTRUCTIONS: ANSWER EACH QUESTION ACCORDING TO THE FOLLOWING KEY:

 (A) Only 1 is correct
 (B) Only 2 is correct
 (C) Both are correct
 (D) Neither are correct

1. Sensory systems are organized in:

 1. series
 2. parallel

ANSWER: C

2. Touch perception:

 1. occurs in the temporal lobe
 2. involves fibers in the fasciculus gracilis

ANSWER: B

3. Convey impulses from touch receptors:

 1. fasciculus gracilis
 2. fasciculus cuneatus

ANSWER: C

4. Touch perception involves the:

 1. medial lemniscal pathway
 2. parietal lobe

ANSWER: C

5. Touch perception involves the:

 1. thalamus
 2. postcentral gyrus of the parietal lobe

ANSWER: C

6. Touch perception involves:

    1. sensory association areas of the postcentral gyrus
    2. the medial lemniscal pathway

ANSWER: C

7. In touch perception the second order neuron:

    1. synapses in the thalamus
    2. crosses to the contralateral side prior to reaching the thalamus

ANSWER: C

8. Pain perception:

    1. occurs in the parietal lobe
    2. is a protective mechanism

ANSWER: C

9. Pain perception:

    1. involves the spinothalamic tract
    2. involves the fasciculus cuneatus

ANSWER: A

10. Pain may be inhibited by:

    1. analgesic drugs
    2. endorphins

ANSWER: C

11. Involved in the perception of sound:

    1. inferior colliculus
    2. medial geniculate nuclei

ANSWER: C

12. Involved in the perception of sound:

    1. thalamus
    2. temporal lobe

ANSWER: C

13. The premotor area:

    1. motor association area
    2. lies just caudal to the primary motor area

ANSWER: A

14. The premotor area:

    1. lies just rostral to the primary motor area
    2. motor association area

ANSWER: C

15. Damage to the pyramidal tract:

    1. paresis of affected muscles
    2. spasticity

ANSWER: C

16. Damage to extrapyramidal tracts:

    1. spastic paralysis
    2. tremors

ANSWER: C

17. Learning:

    1. associative
    2. nonassociative

ANSWER: C

18. Associative learning:

    1. habituation
    2. imprinting

ANSWER: D

19. Associative learning:

    1. insight learning
    2. sensitization

ANSWER: C

20. General interpretative area:

      1. temporal portion is called Wericke's area
      2. damage may result in fluent aphasia

ANSWER: C

INSTRUCTIONS: TRUE / FALSE

21. (T/F): The hypothalamus and the brainstem are responsible for the sleep-wake cycle.

ANSWER: T

22. (T/F): The suprachiasmatic nucleus of the medulla is thought to be the body's biological clock.

ANSWER: F

23. (T/F): Sleep can be induced by injecting serotonin into the preoptic nucleus.

ANSWER: T

24. (T/F): Electrical stimulation of the preoptic nucleus results in inhibition of REM sleep.

ANSWER: F

25. (T/F): Insomnia is a condition characterized by difficulty in sleeping.

ANSWER: T

26. (T/F): An individual who frequently and suddenly lapses into REM sleep, often at inappropriate times, is said to have a condition known as narcolepsy.

ANSWER: T

27. (T/F): Smoking crack cocaine stimulates a massive release of catecholamines in the brain.

ANSWER: T

28. (T/F): Physical dependence on a drug such as heroin is called addiction.

ANSWER: T

29. (T/F): Alcohol is a CNS stimulant.

ANSWER: F

30. (T/F): Alcohol is a CNS depressant.

ANSWER: T

31. (T/F): Extrapyramidal tracts are concerned with gross movements and posture.

ANSWER: T

32. (T/F): Extrapyramidal tracts are concerned with pain and touch perception.

ANSWER: F

33. (T/F): Spastic paralysis may result from damage to the extrapyramidal system.

ANSWER: T

34. (T/F): Habituation is a form of nonassociative learning.

ANSWER: T

35. (T/F): Imprinting is a form of associative learning.

ANSWER: F

36. (T/F): Insight learning is a form of associative learning.

ANSWER: F

37. (T/F): Operant conditioning is a type of associative learning.

ANSWER: T

38. (T/F): Classical conditioning is a type of associative learning.

ANSWER: T

39. (T/F): Information from long-term memory may be retrieved and transferred to short-term memory.

ANSWER: T

40. (T/F): A change in behavior due to experience is defined as memory.

ANSWER: F

41. (T/F): Memory is the storage of knowledge and the ability to recall it.

ANSWER: T

42. (T/F): Paresis is defined as a weakness or slight paralysis of affected muscles.

ANSWER: T

43. (T/F): The premotor area is also known as the motor association area.

ANSWER: T

44. (T/F): The motor system has both serial and parallel organization.

ANSWER: T

45. (T/F): The afferent neurons of the motor system have been referred to as the 'final common pathway.'

ANSWER: F

46. (T/F): Vague awareness of touch is perceived in the medulla.

ANSWER: F

47. (T/F): Vague awareness of touch is perceived in the thalamus.

ANSWER: T

48. (T/F): The primary sensory areas are located in the precentral gyrus.

ANSWER: F

49. (T/F): The primary sensory areas are located in the parietal lobe.

ANSWER: T

50. (T/F): Sensory association areas are also referred to as Brodmann's areas 5 and 7.

ANSWER: T

51. (T/F): Most sensory modalities have only one serial pathway.

ANSWER: F

52. (T/F): Parallel organization is important clinically because when one sensory pathway is damaged, another can function to retain that sense at least to some extent.

ANSWER: T

53. (T/F): Awareness of pain begins in the thalamus.

ANSWER: T

54. (T/F): Substance P is a neurotransmitter and is released by neurons that transmit pain impulses.

ANSWER: T

55. (T/F): Opiate drugs stimulate the release of substance P.

ANSWER: F

56. (T/F): Endorphins and enkephalins inhibit the release of substance P.

ANSWER: T

57. (T/F): Neurons that modify pain sensation are the inhibitory interneurons of the spinal cord.

ANSWER: T

58. (T/F): Fear and anxiety tend to intensify pain.

ANSWER: T

59. (T/F): The perception of pain cannot be influenced by input from higher brain centers.

ANSWER: F

60. (T/F): Perception of pain from an amputated limb is referred to as phantom limb.

ANSWER: T

61. (T/F): Angina pectoris is an example of referred pain.

ANSWER: T

62. (T/F): Brodmann's area 17 is also known as the visual association area.

ANSWER: F

63. (T/F): Brodmann's area 20 is also known as the primary visual area.

ANSWER: F

64. (T/F): Lesions of the visual association areas may result in total blindness.

ANSWER: F

65. (T/F): Lesions of the visual association areas may result in visual agnosia.

ANSWER: T

66. (T/F): Damage to the auditory association areas results in total deafness.

ANSWER: F

67. (T/F): A person who is able to hear sounds but is unable to understand the spoken word may have damage to the auditory association areas.

ANSWER: T

68. (T/F): The pryamidal pathway is conserned with the performance of fine, skilled, learned movements.

ANSWER: T

69. (T/F): In the spinal cord about 50% of the pyramidal axons synapse directly with motor neurons.

ANSWER: F

70. (T/F): The corticospinal tracts are also known as the pyramidal pathway.

ANSWER: T

TEST BANK for CHAPTER 16 : THE PERIPHERAL NERVOUS SYSTEM: AUTONOMIC SYSTEM

INSTRUCTIONS: COMPLETE THE FOLLOWING:

1. The autonomic nervous system uses two neurotransmitters, acetylcholine and:

ANSWER: norepinephrine

2. The exclusive neurotransmitter of the somatic nervous system is:

ANSWER: acetylcholine

3. The autonomic nervous system functions to maintain a steady state within the _____ environment.

ANSWER: internal

4. The efferent component of the autonomic nervous system can be divided into sympathetic and _____ systems.

ANSWER: parasympathetic

5. The _____ component of the ANS dominates during stressful situations.

ANSWER: sympathetic

6. The _____ component of the ANS is most active during periods of emotional calm and physical rest.

ANSWER: parasympathetic

7. The sympathetic system is sometimes referred to as the _____ outflow.

ANSWER: thoracolumbar

8. Cell bodies of preganglionic sympathetic fibers that emerge from the spinal cord are located in the intermedial lateral nuclei of the:

ANSWER: spinal cord

9. White rami communicantes appear white since preganglionic sympathetic neurons are:

ANSWER: myelinated

10. Preganglionic sympathetic neurons secrete the neurotransmitter, _____.

ANSWER: acetylcholine

INSTRUCTIONS: TRUE - FALSE

11. (T/F): Preganglionic sympathetic neurons are designated as cholinergic fibers.

ANSWER: T

12. (T/F): Splanchnic nerves do not contain postganglionic sympathetic fibers.

ANSWER: T

13. (T/F): The adrenal medulla is part of the parasympathetic division of the ANS.

ANSWER: F

14. (T/F): The adrenal medulla is innervated by preganglionic sympathetic fibers.

ANSWER: T

15. (T/F): The parasympathetic division of the ANS is sometimes referred to as the thoracolumbar outflow.

ANSWER: F

16. (T/F): The parasympathetic division of the ANS is sometimes referred to as the craniosacral outflow.

ANSWER: T

17. (T/F): The facial nerve contains parasympathetic fibers.

ANSWER: T

18. (T/F): The oculomotor nerve contains parasympathetic fibers.

ANSWER: T

19. (T/F): The vagus nerve contains parasympathetic and sensory fibers.

ANSWER: T

20. (T/F): The glossopharyngeal nerve contains parasympathetic fibers.

ANSWER: T

21. (T/F): The neurotransmitter utilized by preganglionic fibers of the parasympathetic system is norepinephrine.

ANSWER: F

22. (T/F): Preganglionic and postganglionic fibers of the parasympathetic system utilize acetylcholine as their neurotransmitter.

ANSWER: T

23. (T/F): Following its release, excess acetylcholine is deactivated by the enzyme, cholinesterase.

ANSWER: T

24. (T/F): Muscarinic receptors represent a special type of cholinergic receptor.

ANSWER: T

25. (T/F): Nicotinic receptors are found in the plasma membranes of skeletal muscle fibers.

ANSWER: T

TB.16:4

26. (T/F): Acetylcholine activates muscarinic and nicotinic receptors.

ANSWER: T

27. (T/F): Scopolamine in low doses is utilized to combat motion sickness.

ANSWER: T

28. (T/F): The drug, reserpine, blocks the synthesis and storage of acetylcholine:

ANSWER: F

29. (T/F): Isoproterenol is known as beta blocker.

ANSWER: T

30. (T/F): The drug, pilocarpine, is utilized by ophthalmologists to dilate the iris of the eye.

ANSWER: F

31. (T/F): Stimulation of alpha receptors brings about vasodilatation of blood vessels.

ANSWER: F

32. (T/F): Nicotinic receptors represent a special type of cholinergic receptor.

ANSWER: T

33. (T/F): Many organs are innervated by the sympathetic and parsympathetic divisions of the ANS.

ANSWER: T

34. (T/F): Contraction of a full urinary bladder represents an example of an autonomic reflex integrated within the spinal cord.

ANSWER: T

35. (T/F): The cerebrum and hypothalamus exert strong influences over the ANS in response to stress.

ANSWER: T

36. (T/F): Sympathetic stimulation increases heart rate.

ANSWER: T

37. (T/F): Bronchial tubes are constricted by sympathetic stimulation.

ANSWER: F

38. (T/F): Parasympathetic impulses stimulate secretion by the lacrimal glands.

ANSWER: T

39. (T/F): Pupillary constriction is brought about by parasympathetic stimulation.

ANSWER: T

40. (T/F): Sympathetic stimulation inhibits intestinal motility.

ANSWER: T

41. (T/F): Sympathetic stimulation causes relaxation of the internal sphincter of the urinary bladder.

ANSWER: F

42. (T/F): Parasympathetic stimulation causes the release of fatty acids from adipose tissue.

ANSWER: F

43. (T/F): Each paravertebral sympathetic ganglion chain is contains approximately 22 ganglia.

ANSWER: T

44. (T/F): Gray rami communicantes contain preganglionic sympathetic fibers exclusively.

ANSWER: F

45. (T/F): The celiac ganglia is also referred to as the solar plexus.

ANSWER: T

INSTRUCTIONS: ANSWER EACH QUESTION ACCORDING TO THE FOLLOWING KEY:

    (A) Only 1 is correct
    (B) Only 2 is correct
    (C) Both are correct
    (D) Neither are correct

46. Gray rami communicantes contain:

    1. sensory fibers only
    2. mainly preganglionic parasympathetic fibers

ANSWER: D

47. Regarding epinephrine:

    1. a sympathomimetic drug
    2. acts on alpha receptors

ANSWER: C

48. Regarding epinephrine:

    1. acts on beta receptors
    2. acts on alpha receptors

ANSWER: C

49. Regarding atropine:

    1. an anticholinergic drug
    2. belongs to same class of drugs as does scopolamine

ANSWER: C

50. Regarding the action of reserpine:

   1. inhibits synthesis of epinephrine
   2. inhibits storage of epinephrine

ANSWER: D

51. Nicotinic receptors:

   1. activated by acetylcholine
   2. a type of cholinergic receptor

ANSWER: C

52. Muscarinic receptors:

   1. activated by acetylcholine
   2. a type of cholinergic receptor

ANSWER: C

53. Alpha receptors:

   1. activated by acetylcholine
   2. a type of cholinergic receptor

ANSWER: D

54. Alpha receptors:

   1. activated by epinephrine
   2. a type of adrenergic receptor

ANSWER: C

55. Beta receptors:

   1. activated by epinephrine
   2. a type of adrenergic receptor

ANSWER: C

56. Beta receptors:

   1. activated by norepinephrine
   2. activated by epinephrine

ANSWER: C

57. Beta-blockers:

   1. block action of epinephrine at beta receptors
   2. used to treat cardiac arrhythmias

ANSWER: C

58. Regarding epinephrine:

   1. an adrenergic drug
   2. a sympathomimetic drug

ANSWER: C

59. Stimulation of alpha receptors:

   1. vasoconstriction
   2. intestinal relaxation

ANSWER: C

60. Regarding propranolol:

   1. a beta-blocker
   2. reduces heart rate

ANSWER: C

61. A prevertebral ganglion:

   1. celiac ganglion
   2. superior mesenteric ganglion

ANSWER: C

62. Innervated by the vagus nerves:

   1. stomach
   2. small intestine

ANSWER: C

63. Innervated by the vagus nerves:

   1. heart
   2. large intestine

ANSWER: C

64. Effects of the parasympathetic system:

    1. more localized than those of sympathetic system
    2. longer lasting than those of sympathetic system

ANSWER: A

65. Isoproterenol:

    1. an adrenergic drug
    2. acts directly on beta receptors

ANSWER: C

66. A type of cholinergic receptor:

    1. alpha receptor
    2. beta receptor

ANSWER: D

67. A type of adrenergic receptor:

    1. alpha receptor
    2. beta receptor

ANSWER: C

68. Regarding epinephrine:

    1. activates alpha receptors
    2. activates beta receptors

ANSWER: C

69. Possess cholinergic preganglionic neurons:

    1. sympathetic system
    2. parasympathetic system

ANSWER: C

70. Possess adrenergic postganglionic fibers:

    1. sympathetic system
    2. parasympathetic system

ANSWER: A

TEST BANK for CHAPTER 17: HUMAN SENSES

INSTRUCTIONS: ANSWER EACH QUESTION ACCORDING TO THE FOLLOWING KEY:

(A) Only 1 is correct
(B) Only 2 is correct
(C) Both are correct
(D) Neither are correct

1. The macula lutea:

    1. surrounds the fovea centralis
    2. is typically referred to as the 'white spot'

ANSWER: A

2. The fovea centralis:

    1. is the region of sharpest vision
    2. region where thcones are most concentrated

ANSWER: C

3. Cerumen:

    1. is also referred to as earwax
    2. aids in keeping foreign materials away from the tympanic membrane

ANSWER: C

4. The inner ear:

    1. functions in hearing
    2. functions in equilibrium

ANSWER: C

5. The auricle of the outer ear:

    1. is the rim of the helix
    2. is attached to the skull by muscles and ligaments

ANSWER: B

TB.17:2

6. The tympanic membane:

   1. separates the middle ear from the inner ear
   2. has a diameter of about 3 cm

ANSWER: D

7. A stimulus is:

   1. any detectable change in the environment
   2. detected by the body through its sensory receptors

ANSWER: C

8. Exteroceptors are:

   1. also referred to as visceroceptors
   2. located near the surface of the body

ANSWER: B

9. Proprioceptors provide the CNS with information concerning:

   1. body position
   2. equilibrium

ANSWER: C

10. A gustatory cell may be classified as a:

    1. complex receptor
    2. mechanoreceptor

ANSWER: A

11. The rods of the retina may be classified as:

    1. exteroceptors
    2. simple receptors

ANSWER: A

12. The cones of the retina may be classified as:

    1. enteroceptors
    2. simple receptors

ANSWER: D

13. An olfactory cell may be classified as:

    1. being a complex receptor
    2. a chemoreceptor

ANSWER: B

14. Muscle spindles may be classified as:

    1. complex receptors
    2. proprioceptors

ANSWER: B

15. Exteroceptors of the simple type classification:

    1. Meissner's corpuscles
    2. Merkel's discs

ANSWER: C

16. Enteroceptors of the simple type classification:

    1. Golgi tendon organs
    2. root hair plexuses

ANSWER: D

17. Receptors of the general senses:

    1. Krause's endbulbs
    2. joint kinesthetic receptors

ANSWER: C

18. Receptors exhibiting tonic adaptation:

    1. root hair plexuses
    2. Pacinian corpuscles

ANSWER: D

19. Receptors exhibiting phasic adaptation:

    1. Meissner's corpouscles
    2. free nerve endings (dendrites)

ANSWER: A

20. Receptors located in joint capsules:

   1. Ruffini's end-organs
   2. joint kinesthetic receptors

ANSWER: C

21. Simple receptors may be:

   1. encapsulated
   2. dendrites of sensory neurons

ANSWER: C

22. Receptors may classified according to:

   1. their location in the body
   2. the nature of the stimulus received

ANSWER: C

23. The sensation received by the activation of a particular receptor is called:

   1. modality
   2. an adequate stimulus

ANSWER: A

24. Phasic receptors:

   1. adapt very rapidly
   2. decrease their sensitivity to a given stimulus if the latter is continued for a prolonged period

ANSWER: C

25. Nociceptors respond to:

   1. chemicasl changes
   2. temperature extremes

ANSWER: C

INSTRUCTIONS: TRUE - FALSE

26. (T/F): A sensory unit is represented by a minimum of five neurons and all the receptors they receive input from.

ANSWER: F

27. (T/F): The stimulus that meets lowest threshold for exciting the particular receptor is called the adequate stimulus.

ANSWER: T

28. (T/F): Mechanoreceptors detect touch, vibrations, and pressure.

ANSWER: T

29. (T/F): Somatic sennses have their receptors located in the skin and in muscles, tendons and joints.

ANSWER: T

30. (T/F): Receptors responding to chanmges in electromagnetic energy are called nociceptors.

ANSWER: F

31. (T/F): Tonic receptors adapt less rapidly than do phasic receptors.

ANSWER: T

32. (T/F): Tonic receptors may not possess the ability to adapt to a prolonged stimulus.

ANSWER: T

33. (T/F): The generator potential of a receptor is a local, graded response.

ANSWER: T

34. (T/F): Tactile receptors and nociceptors are often referred to as cutaneous receptors.

ANSWER: T

35. (T/F): Meissner's corpuscles are not very good at detecting two-point discrimination.

ANSWER: F

36. (T/F): The cornea does not contain free nerve endings.

ANSWER: F

37. (T/F): Mechanoreceptors include both stretch and tactile receptors.

ANSWER: T

38. (T/F): Skeletal muscle fibers surrounding the muscle spindles are called intrafusal fibers.

ANSWER: F

39. (T/F): Within muscle spindles, the branches of type II nerve fibers terminate as annulospiral endings.

ANSWER: F

40. (T/F): Olfactory cells are actually bipolar cells.

ANSWER: T

41. (T/F): Absence of the sense of smell is designated an anosmia.

ANSWER: T

42. (T/F): It is believed that different olfactory hairs possess differing types and amounts of receptors.

ANSWER: T

43. (T/F): Taste is the only sense that is conducted directly to the brain without synapsing first in the thalamus.

ANSWER: F

44. (T/F): Foliate papillae do not ordinarily contain taste buds.

ANSWER: F

45. (T/F): Approximately half of the taste buds are known to be associated with circumvallate papillae.

ANSWER: T

46. (T/F): The glossopharyngeal nerves convey sensory information from the anterior third of the tongue.

ANSWER: F

47. (T/F): The space located between the upper and lower eyelids is called the palpebral fissure.

ANSWER: T

48. (T/F): The facial and the vagus nerves each convey information to the brain concerning taste.

ANSWER: T

49. (T/F): The caruncle is located in the lateral canthus.

ANSWER: F

50. (T/F): The conjunctiva secretes mucus to keep the eye moist and lubricated.

ANSWER: T

COMPLETE THE FOLLOWING STATEMENTS

51. The nerve responsible for innervatoing the superior rectus muscle is the:

ANSWER: oculomotor nerve (CN III)

52. The nerve responsible for innervating the superior oblique muscle is the:

ANSWER: trochlear nerve (CN IV)

53. The posterior five-sixths of the fibrous tunic of the eye is made up of the:

ANSWER: sclera

54. The vascular tunic of th eye is also called the:

ANSWER: uvea

55. The most numerous of the photoreceptors in the retina are the:

ANSWER: rods

56. Color vision results from the stimulation of the:

ANSWER: cones

57. The outer layer of the retina is also called the:

ANSWER: pigmented layer

58. The anterior, jagged edge of the nervous layer of the retina is referred to as the:

ANSWER: ora serrata

59. Change of visual focus due to an alteration in the shape of the lens is called:

ANSWER: accomodation

60. Damage to the retina and optic nerve as a result of increased intraocular pressure is called:

ANSWER: glaucoma

61. The bending of light rays upon entering a medium of different density is called:

ANSWE: refraction

62. Irregular curvatures of the cornea or lens results in a condition known as:

ANSWER: astigmatism

63. The chromophore of all visual pigments is termed:

ANSWER: retinal

64. The type of protein in retinal is called an:

ANSWER: opsin

65. Retinal may exist as all-trans retinal or as:

ANSWER: 11-cis retinal

66. The visual pigment of the rods is called:

ANSWER: rhodopsin

67. Rhodopsin is also called:

ANSWER: visual purple

68. Rhodopsin consists of retinal and the opsin known as:

ANSWER: scotopsin

69. Closure of the sodium channels within the rods leads to hyperpolarization and formation of the:

ANSWER: generator potential

70. Transducin activates an enzyme called a:

ANSWER: phosphodiesterase

71. Inflammation of the cornea is called:

ANSWER: keratitis

72. Dark adaptation ordinarily takes about how long to occur?

ANSWER: twenty (20) minutes

73. The most common form of color blindness is:

ANSWER: red-green color blindness

74. Red cones are most sensitive to wavelengths around:

ANSWER: 570 nanometers

75. The color percieved when all of the cones are stimulated is:

ANSWER: white

76. The color perceived when none of the cones are stimulated is:

ANSWER: black

77. Light adaptation usually takes how long to occur?

ANSWER: approximately five (5) minutes

78. The nervous layer of the retina contains two types of cells that help transmit information laterally, the horizontal cells and the:

ANSWER: amacrine cells

79. Axons from the medial aspect of each retina cross to the opposite side as they pass through the:

ANSWER: optic chiasma

80. The optic tracts termiate in the thalamus at the:

ANSWER: lateral geniculate nuclei

INSTRUCTIONS: TRUE - FALSE

81. (T/F): Inflammation of the inner ear is called otitis media.

ANSWER: F

82. (T/F): The middle ear is located within the substance of the parietal bone.

ANSWER: F

83. (T/F): The footplate of the incus rests upon a structure called the oval window.

ANSWER: F

84. (T/F): When the tensor tympani contracts it pulls the malleus medially.

ANSWER: T

85. (T/F): The spaces of the bony labyrinth are filled with a fluid called perilymph.

ANSWER: T

86. (T/F): The utricle and saccule contain specialized areas that function in equlibrium called the maculae.

ANSWER: T

87. (T/F): The cupula is a gelatinous substance in which the hairs of hair cels are embedded.

ANSWER: T

88. (T/F): Groups of sensitive hair cells located on the ampullary crests are called maculae.

ANSWER: F

89. (T/F): The central bony core of the cochlea is called the modiolus.

ANSWER: T

90. (T/F): The organs of Corti are also referred to as the spiral organs.

ANSWER: T

91. (T/F): The flexible gelatinous flap that extends over the organs of Corti is called the tectorial membrane.

ANSWER: T

92. (T/F): The distance between the peaks and valleys of a soundwave is called its amplitude.

ANSWER: T

93. (T/F): The distance between the peaks and valleys of a soundwave is called its wavelength.

ANSWER: F

94. (T/F): The number of wavelengths occurring per unit time is called the grequency.

ANSWER: T

95. (T/F): The wavelength of of a sound is experienced as pitch.

ANSWER: F

96. (T/F): The function of the outer and middle ear is to transmit sound waves.

ANSWER: T

97. (T/F): Transmission of soundwaves from the tympanic membrane to the oval window results in a very large increase in the force transmitted per unit area.

ANSWER: T

98. (T/F): Pressure waves transmitted to the scala tympani are dissipated by the oval window.

ANSWER: F

99. (T/F): High-frequency sounds cause the distal end of the basilar membrane to vibrate more than other parts.

ANSWER: F

100. (T/F): Prolonged exposure to high-intensity sound can cause damage to the hair cells of the organs of Corti.

ANSWER: T

101. (T/F): Orientation of the body relative to the ground is called static equlibrium.

ANSWER: T

102. (T/F):   The otolithic membrane is believed to be produced by
              the supporting cells of the maculae.

ANSWER: T

103. (T/F):   The otolithic membrane contains small particles of
              potassium carbonate called otoliths.

ANSWER: F

104. (T/F):   The optic vesicles develop from the region of the
              primitive brain called the diencephalon.

ANSWER: T

105. (T/F):   Angular movement causes the endolymph in the
              semicircular ducts to move in a direction opposite the
              rotation.

ANSWER: T

106. (T/F):   The ampullae of the semicircular canals can detect
              movement in three-dimensional space.

ANSWER: T

107. (T/F):   The middle ear cavity and the eustachian tube develop
              from the embryonic first pharyngeal pouch.

ANSWER: T

108. (T/F):   The otic placode develops as a thickening of surface
              ectoderm.

ANSWER: T

109. (T/F):   The lens placodes of the developing eye arise from
              endoderm.

ANSWER: F

TB.18:1

TEST BANK for CHAPTER 18: ENDOCRINE CONTROL

INSTRUCTIONS: TRUE - FALSE

1. (T/F):   The study of endocrine function and malfunction is called endocrinology.

ANSWER: T

2. (T/F):   Endocrine glands are classified as glands having a multitude of different ducts.

ANSWER: F

3. (T/F):   Chemical messengers secreted by endocrine glands are called hormones.

ABSWER: T

4. (T/F):   The tissues upon which hormones exert their influence are called 'target tissues.

ANSWER: T

5. (T/F):   The hypothalamus contains neurosecretory cells.

ANSWER: T

6. (T/F):   Melatonin is secreted by the pineal gland.

ANSWER: T

7. (T/F):   Melatonin may help control the onset of puberty in humans.

ANSWER: T

8. (T/F):   Calcitonin elevates blood-calcium level by stimulating osteoclasts to breakdown bone.

ANSWER: F

9. (T/F): Hormones are considered to be chemically diverse substances.

ANSWER: T

10. (T/F): Oxytocin is an example of a steroid hormone.

ANSWER: F

11. (T/F): Antidiuretic hormone (ADH) is an example of a hormone derived from a fatty acid.

ANSWER: F

12. (T/F): Progesterone is a steroid hormone from which several other hormones are generated.

ANSWER: T

13. (T/F): The effect of one hormone cannot be enhanced by the presence of a second hormone.

ANSWER: F

14. (T/F): Hormones may act in a synergistic manner.

ANSWER: T

15. (T/F): Many hormones act through second messengers such as cyclic AMP.

ANSWER: T

16. (T/F): Prostaglandins are examples of local chemical mediators.

ANSWER: T

17. (T/F): A calcium-calmodulin complex may activate certain enzymes.

ANSWER: T

18. (T/F): Prostaglandins are only released by blood platelets.

ANSWER: F

19. (T/F): Prostaglandins are derived from fatty acids.

ANSWER: T

20. (T/F): Prostaglandins are frequently referred to as local hormones.

ANSWER: T

21. (T/F): Hyposecretion of growth hormone may lead to a condition called pituitary dwarfism.

ANSWER: T

22. (T/F): In children, hyposecretion of thyroid hormones may lead to a condition called cretinism.

ANSWER: T

23. (T/F): Hypoglycemia may result in circumstances in which there is a hyposecretion of insulin.

ANSWER: F

24. (T/F): Chushing's disease may result from hyposecretion of adrenocortical hormones.

ANSWER: F

25. (T/F): Addison's disease typically results from a hyposecretion of growth hormone in children.

ANSWER: F

26. (T/F): Tetany and death may result from hypersecretion of parathyroid hormone.

ANSWER: T

27. (T/F): A common disorder associated with hyposecretion of insulin is diabetes mellitus.

ANSWER: T

28. (T/F):   Increased metabolic rate, nervousness and irritability may result from hyposecretion of thyroid hormones.

ANSWER: F

29. (T/F):   Patients with Addison's disease are unable to synthesize sufficient glucose by gluconeogenesis.

ANSWER: T

30. (T/F):   Enlargement of the thyroid gland is clinically known as goiter.

ANSWER: T

INSTRUCTIONS: ANSWER EACH QUESTION ACCORDING TO THE FOLLOWING KEY:

    (A) Only 1 is correct
    (B) Only 2 is correct
    (C) Both are correct
    (D) Neither are correct

31. The posterior lobe of the pituitary gland

    1. is called the neurohypophysis
    2. develops from the hypothalamus

ANSWER: C

32. Antidiuretic hormone

    1. is synthesized by cells of the neurohypophysis
    2. is secreted by pituicytes

ANSWER: D

33. Antidiuretic hormone is:

    1. also called vassopressin
    2. actually synthesized by cells of the hypothalamus

ANSWER: C

34. The hormone, oxytocin:

      1. stimulates contraction of uterine smooth muscle
      2. inhibits the release of milk from the breast

ANSWER: A

35. Pituicytes of the neurohypophysis:

      1. are supporting cells
      2. secrete oxytocin

ANSWER: A

36. Osmoreceptors in the hypothalamus:

      1. may trigger ADH synthesis in cells of the neurohypophysis
      2. may inhibit the synthesis of ADH when the blood becomes too dilute

ANSWER: B

37. The condition known as diabetes insipidus:

      1. may result from damage to ADH-producing cells
      2. is characterized by an inability to effectively regulate urine volume

ANSWER: T

38. The anterior lobe of the pituitary gland:

      1. is also called the neurohypophysis
      2. secretes at least six hormones

ANSWER: B

39. Secreted by the adenohypophysis:

      1. prolactin
      2. growth hormone

ANSWER: C

40. Secreted by the anterior lobe of the pituitary gland:

   1. oxytocin
   2. growth hormonw

ANSWER: B

41. Secreted by the adenohypohysis:

   1. beta-lipoprotein
   2. precursor of endorphins and enkephalins

ANSWER: C

42. Somatotrophin:

   1. is also called growth hormone
   2. stimulates protein synthesis

ANSWER: T

43. Somatomedins:

   1. are synthesized in the liver
   2. promote linear growth of the skelton

ANSWER: C

44. Some hormones of the anterior lobe of the pituitary may be regulated by a:

   1. releasing factor
   2. inhibiting factor

ANSWER: C

45. Growth hormone secretion is:

   1. inhibited by hyperglycemia
   2. stimulated by somatostatin

ANSWER: A

46. Acromegaly is a clinical condition in which:

   1. there is a hypersecretion of growth hormone
   2. bones of the hands, feet and face increase in diamete

ANSWER: T

47. Milk release from the breast:

   1. is stimulated by the infant sucking at the breast
   2. is a direct result of prolactin secretion

ANSWER: A

48. The hormone, prolactin stimulates:

   1. milk production by cells of the mammary glands
   2. milk relase from the breasts following a sucking stimulus by the infant

ANSWER: A

49. Secreted by the anterior lobe of the pituitary gland:

   1. TSH
   2. FSH

ANSWER: C

50. Secreted by the anterior lobe of the pituitary gland:

   1. ACTH
   2. ADH

ANSWER: A

COMPLETE THE FOLLOWING STATEMENTS

51. Hormones that stimulate the activity of other endocrine glands are referred to as:

ANSWER: tropic hormones

52. Follicle cells of the thyroid gland secrete a protein called:

ANSWER: thyroglobulin

53. Thyroglobulin is stored within a protein substance called the:

ANSWER: colloid

54. Grave's disease is a clinical condition involving an increase in size and activity of the:

ANSWER: thyroid gland

55. Parathyroid hormone is secreted by cells called:

ANSWER: chief cells

56. Parathyroid hormone can indirectly increase the amount of calcium absorbed from the intestine by activating:

ANSWER: vitamin D

57. Substances that inhibit iodine uptake or effectiveness are called:

ANSWER: goitrogens

58. Any abnormal enlargement of the thyroid gland is termed a:

ANSWER: goiter

59. Hyperparathyroidism is often caused by small benign tumors called:

ANSWER: adenomas

60. The symptoms of hypoparathyroidism may be relieved by the injection of PTH or:

ANSWER: calcium

61. When calcium levels become excessive the thyroid gland releases a hormone called:

ANSWER: calcitonin

62. Alpha cells of the pancreatic islets secrete a hormone called:

ANSWER: glucagon

63. The delta cells of the pancreatic islets secrete a hormone called:

ANSWER: somatostatin or (GHIH)

64. Approximately 70% of the pancreatic islet cells are called:

ANSWER: beta cells

65. Liver cells convert glycogen to glucose by a process called:

ANSWER: glycogenolysis

66. Liver cells may be stimulated to produce glucose from noncarbohydrates, a process called:

ANSWER: gluconeogenesis

67. Glucagon's principal function is to elevate the blood-:

ANSWER: sugar level

68. In the absence of insulin, increased fat metabolism leads to an increase in the formation of:

ANSWER: ketone bodies

69. The build up of ketone bodies in the blood is called:

ANSWER: ketosis

70. Increased urine volume is clinically referred to as:

ANSWER: polyuria

71. Constant thirst is clinically termed:

ANSWER: polydipsia

72. Approximately 80% of the hormone output of the adrenal medulla is:

ANSWER: epinephrine

73. Hormones that mimic the actions of those produced by the sympathetic nervous system are classified as:

ANSWER: sympathomimetic

74. Mineralocorticoids are produced by the region of the adrenal cortex called the:

ANSWER: zona glomerulosa

75. The region of the adrenal cortex primarily responsible for the production of sex hormones is the:

ANSWER: zona reticularis

76. The middle region of the adrenal cortex is called the:

ANSWER: zona fasiculata

77. Male sex hormones are also termed:

ANSWER: androgens

78. The principal glucocorticoid activity of the adrenal cortex may be accounted for by the hormone:

ANSWER: cortisol (hydrocortisone)

79. The hormones released by the pineal gland include melatonin and:

ANSWER: adrenoglomerulotropin

80. In response to increased blood volume, stretched cardiac muscle fibers release a hormone called:

ANSWER: atrial natriuretic factor (ANF)

TB.19:1

TEST BANK for CHAPTER 19: THE BLOOD

INSTRUCTIONS: SELECT THE ONE BEST ANSWER

1. The normal circulating blood volume represents approximately what percentage of one's body weight?

a) 8%
b) 18%
c) 28%
d) 35%

ANSWER: a

2. The range of the normal hematocrit in the adult male is:

a) 36% - 46%
b) 42% - 54%
c) 52% - 67%
d) 64% - 78%

ANSWER: b

3. The range of the normal hematocrit in the adult female is:

a) 36% - 46%
b) 42% - 54%
c) 52% - 67%
d) 64% - 78%

ANSWER: a

4. The number of red blood cells per cubic millimeter of blood in an adult male is approximately:

a) 2.5 million
b) 5.4 million
c) 2.5 billion
d) 5.4 million

ANSWER: b

5. The diameter of a normal red blood cell is approximately:

a) 2.5 micrometers
b) 7.5 micrometers
c) 10.5 micrometers
d) 15.5 micrometers

ANSWER: b

6. When blood is lost due to hemorrhage, the amount of time required for the plasma level to return to normal is typically:

a) 24 hours
b) 2 - 3 days
c) 7 - 8 days
d) a minimum of 2 weeks

ANSWER: b

7. Under normal conditions, eosinophils constitute what percentage of the total WBC population?

a) 1% - 3%
b) 6% - 8%
c) 10% - 12%
d) less than 1%

ANSWER: a

8. Under normal conditions, lymphocytes constitute what percentage of the total WBC poulation?

a) 1% - 3%
b) 6% - 10%
c) 25% - 35%
d) 40% - 55%

ANSWER: c

9. Under normal conditions, monocytes constitute what percentage of the total WBC population?

a) less than 1%
b) 1% - 3%
c) 5% - 6%
d) 10% - 15%

ANSWER: c

10. Under normal conditions, neutrophils constitute what percentage of the total WBC population?

a) 10%
b) 20%
c) 40%
d) 60%

ANSWER: d

11. The largest cells of the WBC population are the:

a) neutrophils
b) eosinophils
c) monocytes
d) basophils

ANSWER: c

12. The number of platelets per cubic millimeter of circulating blood in a healthy individual is approximately:

a) 10,000
b) 50,000
c) 100,000
d) 300,000

ANSWER: d

13. Fibrinogen is also known as:

a) Factor I
b) Factor II
c) Factor III
d) Factor X

ANSWER: a

14. Factor IV is also known as:

a) phosphorous
b) iodine
c) calcium
d) sodium

ANSWER: c

15. Antihemophilic factor is also known as:

a) Factor III
b) Factor V
c) Factor VIII
d) Factor IX

ANSWER: c

16. Hageman factor is also known as:

a) Factor I
b) Factor XI
c) Factor XII
d) Factor XIII

ANSWER: c

17. White blood cells are outnumbered by red blood cells almost:

a) 10 to 1
b) 100 to 1
c) 700 to 1
d) 1,000 to 1

ANSWER: c

18. The average circulating life span of a red blood cell is about:

a) 10 days
b) 50 days
c) 75 days
d) 120 days

ANSWER: d

19. In normal blood the concentration of carboxyhemoglobin is approximately:

a) 0.5%
b) 5.0%
c) 10%
d) greater than 10%

ANSWER: a

20. Each molecule of hemoglobin consists of:

a) one alpha chain and one beta chain
b) two alpha chains and two beta chains
c) one alpha chain and two beta chains
d) two alpha chains and one beta chain

ANSWER: b

INSTRUCTIONS: ANSWER EACH QUESTION ACCORDING TO THE FOLLOWING KEY:

(A) Only 1 is correct
(B) Only 2 is correct
(C) Both are correct
(D) Neither are correct

21. Red blood cells:

1. have a circulating life span of about 120 days
2. are shaped like biconcave discs

ANSWER: C

TB.19:5

22. Eosinophils are white blood cells that:

       1. constitute about 1 - 5% of the total WBC population
       2. are categorized as agranulocytes

ANSWER: A

23. Agranular leukocytes include the:

       1. monocytes
       2. lymphocytes

ANSWER: C

24. Agranular leukocytes include the:

       1. basophils
       2. neutrophils

ANSWER: D

25. Granular leukocytes include the:

       1. lymphocytes
       2. monocytes

ANSWER: D

26. Granular leukocytes include the:

       1. eosinophils
       2. neutrophils

ANSWER: C

27. Leukopenia may be present in:

       1. viral infections
       2. cirrhosis

ANSWER: C

28. Leukopenia may be present in:

       1. rheumatoid arthritis
       2. viral infections

ANSWER: C

29. Leukemia may:

       1. be of an acute or chronic nature
       2. occur at any age

ANSWER: C

30. Exposure to which of the following has been linked to the occurrence of leukemia?

   1. radiation
   2. benzene

ANSWER: C

31. Platelets:

   1. are incomplete cells
   2. arise from monocytes

ANSWER: A

32. Platelets:

   1. are also called thrombocytes
   2. arise from megakaryocytes

ANSWER: C

33. Platelets:

   1. function in the process of hemostasis
   2. are multinucleated

ANSWER: A

34. Leukemia:

   1. may lead to the development of anemia
   2. may result in impaired blood clotting

ANSWER: C

35. Platelets function in the process of hemostasis by:

   1. forming physical plugs in a broken wall of a blood vessel
   2. releasing chemicals that promote the clotting process

ANSWER: C

36. The basophils:

   1. account for less than 1% of the circulating WBCs
   2. may be involved in fat metabolism

ANSWER: C

37. The basophils:

   1. contain sparse amounts of histamine
   2. contain heparin

ANSWER: B

38. The monocytes:

   1. are the smallest of the WBCs
   2. develop into macrophages after leaving the circulation

ANSWER: B

39. The eosinophils:

   1. contain large amounts of histamine
   2. are typically more numerous than lymphocytes

ANSWER: D

40. Neutrophils:

   1. account for about 60% of all leukocytes
   2. exist largely in the band form in the circulating blood

ANSWER: A

INSTRUCTIONS: TRUE - FALSE

41. (T/F): Death in leukemia patients often results from bacterial infection?

ANSWER: T

42. (T/F): Megakaryocytes give rise to platelets primarily within the liver:

ANSWER: F

43. (T/F): Leukopenia is a condition which is typically indicative of a decreased number of circulating neutrophils.

ANSWER: T

44. (T/F): Aspirin has been shown to inhibit platelet aggregation by inhibiting the synthesis of prostaglandins.

ANSWER: T

45. (T/F): Red blood cells and white blood cells enter the circulation by a process called diapedesis.

ANSWER: T

46. (T/F): Erythropoietin is a hormone that is secreted by the kidneys.

ANSWER: T

47. (T/F): The hormone, erythropoietin, is synthesized by the liver and kidneys.

ANSWER: T

48. (T/F): Erythropoietin is a hormone that lengthens the time required for red blood cell maturation.

ANSWER: F

49. (T/F): An increased rate of red blood cell destruction may lead to a condition called anemia.

ANSWER: T

50. (T/F): Pernicious anemia may result from a deficiency of intrinsic factor.

ANSWER: T

51. (T/F): Hyperchromic red blood cells typically represent large, mature red blood cells.

ANSWER: F

52. (T/F): Erythropoietin plays an important homeostatic role in maintaining normal numbers of RBCs and an adequate oxygen supply.

ANSWER: T

53. (T/F): Carboxyhemoglobin has a half-life of approximately 24 hours.

ANSWER: F

54. (T/F): Cigarette smokers typically have a lower carboxyhemoglobin concentration in their blood than do nonsmokers.

ANSWER: F

55. (T/F): An increased incidence of atherosclerosis is associated with increased levels of carboxyhemoglobin in the blood.

ANSWER: T

56. (T/F): Ten to fourteen days are typically required for a stem cell to give rise to a fully differentiated mature erythrocyte.

ANSWER: F

57. (T/F): Reduced hemoglobin tends to appear bright red in color.

ANSWER: F

58. (T/F): Methemoglobin has a very strong affinity for binding oxygen.

ANSWER: F

59. (T/F): Cyanosis is indicative of the presence of large quantities of oxyhemoglobin within vessels of the skin.

ANSWER: F

60. (T/F): A detached blood clot is clinically termed a thrombus.

ANSWER: F

COMPLETE THE FOLLOWING STATEMENTS

61. During the clotting process, prothrobin activator catalyzes the conversion of prothrobin to its active form,:

ANSWER: thrombin

62. Inappropriate clotting may occur when blood flow is sluggish, a condition referred to as:

ANSWER: stasis

63. Clotting that occurs in an unbroken blood vessel is called a:

ANSWER: thrombosis

64. Most persons of western European descent are Rh:

ANSWER: positive

65. Persons with type O blood are sometimes referred to as universal:

ANSWER: donors

66. Persons with type AB blood are referred to as universal:

ANSWER: recipients

67. Heparin inactivates several clotting factors and is thus a potent:

ANSWER: anticoagulant

68. The enzyme activated by a combination of thrombin, activated factor XII, and lysosomal enzymes from damaged tissues is called:

ANSWER: plasmin

69. The most common form of maternal-fetal blood type incompatibility is called:

ANSWER: erythroblastosis fetalis

70. The most common Rh antigen in persons of western European descent is antigen:

ANSWER: D

71. Approximately 15% of persons of western European descent are Rh:

ANSWER: negative

72. Agglutination occurs because red blood cells have specific proteins on their surface called antigens, or:

ANSWER: agglutinogens

73. Antibodies, present in the plasma, that cause agglutination are called:

ANSWER: agglutinins

74. The ion required in order for prothrombin activator to convert prothrombin to thrombin is:

ANSWER: calcium

75. Thrombin acts as an enzyme to convert plasma fibrinogen to:

ANSWER: fibrin

76. The clotting factor required to convert loose fibrin to dense fiobrin is:

ANSWER: factor XIII

77. "Bleeder's disease" is also termed:

ANSWER: hemophilia

78. The process of blood clotting is also called:

ANSWER: coagulation

79. Factor XII clotting factor is also known as:

ANSWER: Hageman factor

80. As a neutrophil matures its nucleus becomes indented or horseshoe shaped and at this stage of development the cell is called a band or:

ANSWER: stab neutrophil

81. Monocytes that migrate from the circulation into the surrounding tissues develop into:

ANSWER: macrophages

82. A form of cancer in which any one of the white blood cells proliferates wildly within the bone marrow is called:

ANSWER: leukemia

83. The primary function of platelets is to stop bleeding, a process termed:

ANSWER: hemostasis

84. The proportional representation of each kind of blood cell is typically reported clinically in a:

ANSWER: differential blood cell count

85. Living at a very high altitude may lead to the development of a homeostatic condition known as secondary:

ANSWER: polycythemia

86. A specific type of anemia in which red blood cells are simply not manufactured is called:

ANSWER: aplastic anemia

87. Hemoglobin combines weakly with oxygen to form:

ANSWER: oxyhemoglobin

88. Hemoglobin which is not combined with oxygen is called:

ANSWER: reduced hemoglobin

89. In its ferric form, hemoglobin is referred to as:

ANSWER: methemoglobin

90. Methemoglobin imparts a bluish color to the skin, a condition called:

ANSWER: cyanosis

91. Once a devloping red blood cell expells its nucleus it is called a:

ANSWER: reticulocyte.

92. The process by which reticulocytes squeeze through the intact walls of capillaries is called:

ANSWER: diapedesis

93. Normal plasma clotting time is approximately 11 seconds and is also called the:

ANSWER: prothrombin time (PT)

94. Red blood cells which have aggregated like a stack of coins are referred to as:

ANSWER: rouleaux

95. The destruction of red blood cells and the subsequent release of hemoglobin is referred to as:

ANSWER: hemolysis

TEST BANK for CHAPTER 20: THE HEART

INSTRUCTIONS: ANSWER EACH QUESTION ACCORDING TO THE FOLLOWING KEY:

    (A) Only 1 is correct
    (B) Only 2 is correct
    (C) Both are correct
    (D) Neither are correct

1. The heart:

    1. is located within the mediastinum
    2. has its base directed posterosuperiorly

ANSWER: C

2. The apex of the heart:

    1. lies to the left of the body's midline
    2. is typically located in the seventh intercostal space

ANSWER: A

3. The apex of the heart:

    1. is composed largely of the tip of the right ventricle
    2. is directed anteroinferiorly and to the right

ANSWER: D

4. The pericardium:

    1. surrounds the heart
    2. encloses the origin of the pulmonary trunk

ANSWER: C

5. The pericardium:

    1. has an outer parietal component
    2. has an inner layer called the visceral pericardium

ANSWER: C

6. The epicardium:

    1. is the outer layer of the heart wall
    2. is formed by the visceral pericardium

ANSWER: C

7. The pericardial cavity:

   1. is a potential space
   2. typically contains about 25 to 35 milliliters of fluid

ANSWER: C

8. The pericardial cavity:

   1. lies between the visceral and parietal pericardium
   2. is a serous cavity

ANSWER: C

9. The fibrous layer of the parietal pericardium:

   1. fuses with the diaphragm
   2. is composed largely of mesothelium

ANSWER: A

10. Regarding the heart wall:

    1. the inner layer is called the epicardium
    2. the middle layer is called the myocardium

ANSWER: B

11. The endocardium:

    1. consists largely of an endothelial lining
    2. is continuous with the endothelial lining
       of the great vessels entering and leaving the heart

ANSWER: C

12. The myocardium:

    1. is of uniform thickness throughout the heart
    2. consists primarily of cardiac muscle

ANSWER: B

13. The myocardium:

    1. of the left ventricle is thicker than that of the right
    2. lies between the endocardium and epicardium

ANSWER: C

14. The connective tissue 'skeleton' of the heart:

    1. divides the atria from the ventricles
    2. divides the heart into right and left halves

ANSWER: C

15. The coronary sulcus:

    1. is observed only on the internal aspect of the heart
    2. is indicative of the separation between the atria and ventricles

ANSWER: B

16. The pulmonary trunk:

    1. carries blood from the left ventricle to the lungs
    2. lies anterior to the origin of the left coronary artery

ANSWER: B

17. The right and left atria:

    1. each possess an auricle
    2. each possess musculi pectinati

ANSWER: C

18. The superior vena cava:

    1. opens into the right atrium
    2. is typically located to the right of the ascending aorta

ANSWER: C

19. Cor triloculare biatriatum:

    1. is a congenital absence of the interventricular septum
    2. is also called the 'Swiss cheese" form of VSD

ANSWER: A

20. The most common form of ASD:

    1. is the complete absence of an interatrial septum
    2. is typically the result of a patent foramen ovale

ANSWER: B

21. The venae cordis minimae:

    1. return most of the blood of the heart wall to the right atrium
    2. empty their blood into all four chambers of the heart

ANSWER: B

22. The right and left venticles of the heart:

    1. exhibit trabeculae carnae
    2. exhibit chordae tendineae

ANSWER: C

23. Mitral stenosis:

    1. impedes the flow of blood from the left atrium into the left ventricle
    2. results in an increased pressure in the left atrium

ANSWER: C

24. Mitral valve insufficiency may lead to:

    1. pulmonary edema
    2. a decrease in cardiac output

25. Mitral valve insufficiency may lead to:

    1. left ventricular hypertrophy
    2. increased left atrial pressure

ANSWER: C

COMPLETE THE FOLLOWING STATEMENTS

26. Thickening of the mitral valve with resultant narrowing of the valve orifice is clinically termed:

ANSWER: mitral stenosis

27. When a valve cusp is shortened or by some other means prevented from closing completely, the condition is referred to clinically as:

ANSWER: valvular insufficiency

28. The bicuspid, or left atrioventricular valve, is also called the:

ANSWER: mitral valve

29. The cusps of the right AV valve are designated as anterior, posterior and:

ANSWER: septal

30. Major branches of the right coronary artery include the AV nodal, posterior interventricular and:

ANSWER: marginal

31. Major branches of the left coronary artery include the anterior interventricular and:

ANSWER: circumflex

32. The great cardiac vein, middle cardiac vein, and small cardiac vein are all tributaries of the:

ANSWER: coronary sinus

33. The coronary sinus empties its blood into the:

ANSWER: right atrium

34. The sinoatrial (SA) node is frequently referred to as the heart's:

ANSWER: pacemaker

35. The sinoatrial (SA) node is located within the superior aspect of the:

ANSWER: right atrium

36. The action potential generated by the sinoatrial (SA) node is conducted to the atrioventricular (AV) node by means of:

ANSWER: internodal fiber bundles

37. The atrioventricular (AV) node is located within the right atrium near the lower portion of the:

ANSWER: interatrial septum

38. Once the action potential leaves the AV node, it enters specialized muscle fibers called:

ANSWER: Purkinje fibers

39. Purkinje fibers are grouped into a collective mass termed the:

ANSWER: atrioventricular (AV) bundle (of His)

40. The AV bundle ultimately divides into right and left:

ANSWER: bundle branches

INSTRUCTIONS: TRUE - FALSE

41. (T/F): Purkinje fibers have a conduction velocity that is about six times that of ordinary cardiac muscle.

ANSWER: T

42. (T/F): There is normally a short delay at the AV node before the action potential is transmitted to the ventricles.

ANSWER: T

43. (T/F): Intercalated discs are specialized cell junctions that offer substantial resistance to the passage of an action potential from one cell to another.

ANSWER: F

44. (T/F): Cardiac muscle is frequently referred to as a 'functional syncytium.'

ANSWER: T

45. (T/F): The presence of intercalated discs permits cardiac muscle to be referred to as a 'functional syncytium.'

ANSWER: T

46. (T/F): In a normal individual the cardiac cycle typically lasts for a period of about 0.8 second.

ANSWER: T

47. (T/F): The first heart sound results from the closure of the pulmonic and aortic valves.

ANSWER: F

48. (T/F): Cardiac output represents that volume of blood pumped by one ventricle in one minute.

ANSWER: T

49. (T/F): Stroke volume is regulated primarily by heart rate and parasympathetic stimulation.

ANSWER: F

50. (T/F): Ventricular escape is the result of new rhythmic impulses being generated in a site other than the SA node.

ANSWER: T

51. (T/F): Stimulation of vagal fibers to the heart results in the release of the neurotransmitter substance called epinephrine.

ANSWER: F

52. (T/F): A resting heart rate in excess of 100 beats per minute represents a condition called bradycardia.

ANSWER: F

53. (T/F): Any alteration in the normal rhythm of the heart beat is termed an arrhythmia.

ANSWER: T

54. (T/F): Should the PR interval exceed 0.20 second in a person with a normal heart rate, the individual is considered to have first-degree heart block.

ANSWER: T

55. (T/F): An ECG pattern exhibiting 'dropped' QRS complexes is characteristic of a condition referred to as second-degree heart block.

ANSWER: T

56. (T/F): Cardiac fibrillation may result from a deficiency of sodium ions in the extracellular environment.

ANSWER: T

57. (T/F): An excess of potassium ions in the extracellular environment markedly increases heart rate as well as the strength of contraction.

ANSWER: F

58. (T/F): The secretion of norepinephrine leads to an increased heart rate and strength of contraction.

ANSWER: T

59. (T/F): The infant heart rate is typically faster than that of a child or an adult.

ANSWER: T

60. (T/F): Sympathetic stimulation of the heart results in a decrease in heart rate.

ANSWER: F

INSTRUCTIONS: SELECT THE ONE BEST ANSWER

61. The infant heart rate is typically:

a) 50 beats per minute
b) 60 beats per minute
c) 80 beats per minute
d) 120 beats per minute

ANSWER: d

62. Internodal fiber bundles have a conduction velocity of approximately:

a) 1.0 meter per second
b) 3.0 meters per second
c) 5.0 meters per second
d) 6.0 meters per second

ANSWER: a

63. Bradycardia is typically employed to designate a heart rate of:

a) less than 60 beats per minute
b) less than 72 beats per minute
c) greater than 80 beats per minute
d) greater than 100 beats per minute

ANSWER: a

64. The normal QT interval typically lasts for about:

a) 0.10 second
b) 0.35 second
c) 0.50 second
d) 0.75 second

ANSWER: b

65. The normal PR interval typically lasts for about:

a) 0.16 second
b) 0.26 second
c) 0.36 second
d) 0.46 second

ANSWER: a

66. Known to increase heart rate:

a) excess potassium ions in extracellular environment
b) parasympathetic stimulation
c) thyroxine
d) hypothermia

ANSWER: c

67. The pulmonary veins:

a) convey oxygen poor blood
b) empty into the right atrium
c) are tributaries of the coronary sinus
d) are typically four in number

ANSWER: d

68. Which is NOT a feature of the left atrium?

a) musculi pectinati
b) auricle
c) openings for vena cordis minimae
d) trabeculae carnae

ANSWER: d

69. Which is NOT a feature of the right ventricle?

a) trabeculae carnae
b) papillary muscle
c) musculi pectinati
d) openings for venae cordis minimae

ANSWER: c

70. Which is NOT a risk factor in accentuating the development of atherosclerosis?

a) cigarette smoking
b) obesity
c) high dietary fat intake
d) high dietary protein intake

ANSWER: d

TEST BANK for CHAPTER 21: CIRCULATION: THE BLOOD VESSELS

INSTRUCTIONS: COMPLETE THE FOLLOWING:

1. The principal types of blood vessels include arteries, capillaries, and:

ANSWER: veins

2. From arterioles, blood flows into:

ANSWER: capillaries

3. Blood typically passes from capillaries into:

ANSWER: venules

4. Venules unite to form larger vessels called:

ANSWER: veins

5. The wall of an artery or vein has how many layers?

ANSWER: three

6. The tunic intima consists of a basement membrane and a lining of:

ANSWER: endothelial cells

7. The tunica media consists of elastic connective tissue and circular:

ANSWER: smooth muscle cells

8. Smooth muscle cells of the tunica media are innervated by which division of the autonomic nervous system?

ANSWER: sympathetic

9. The outermost wall of an artery or vein is called the tunica:

ANSWER: adventitia

10. In large veins, the thickest layer of the wall is the tunica:

ANSWER: adventitia

11. Tiny blood vessels in the walls of larger arteries and veins are called the:

ANSWER: vasa vasorum

12. Blood within the arteries is referred to as:

ANSWER: arterial blood

13. The largest of the arteries in the body are the:

ANSWER: elastic arteries

14. The three main types of arteries are muscular arteries, elastic arteries, and:

ANSWER: arterioles

15. The pulmonary trunk is an example of what type of artery?

ANSWER: elastic artery

16. Elastic arteries are also referred to as the:

ANSWER: conducting arteries

17. Muscular arteries are also referred to as the:

ANSWER: distributing arteries

18. The axillary, brachial, and femoral arteries are examples of which type of artery?

ANSWER: muscular

19. From the distributing arteries, blood flows into the:

ANSWER: arterioles

20. The diameter of an arteriole averages about 30:

ANSWER: micrometers

21. The average diameter of a typical muscular artery is approcximately 0.4:

ANSWER: centimeter

22. Sympathetic innervation of vascular smooth muscle causes the muscles to:

ANSWER: contract

23. Contraction of vascular smooth muscle cells results in a narrowing of the vessel called:

ANSWER: vasoconstriction

24. As vascular smooth muscle relaxes, the vessel enlarges in diameter, a process called:

ANSWER: vasodilation

25. Increased blood flow to a tissue in response to increased metabolic activity is called:

ANSWER: active hyperemia

26. Capillary beds typically begin with an extension of the arteriole called the:

ANSWER: metarteriole

27. Blood flow into capillaries is regulated by the contraction and relaxation of structures called:

ANSWER: precapillary sphincters

28. The distal portion of the metarteriole is referred to as the:

ANSWER: thoroughfare channel

29. The sites of exchanges between the blood and the tissue cells are called:

ANSWER: true capillaries

30. The two basic types of capillaries in the body are the continuous capillary and the:

ANSWER: fenestrated capillary

31. In the liver, arterioles and venules are interconnected by modified capillaries called:

ANSWER: sinusoids

32. Blood within the veins is referred to as:

ANSWER: venous blood

33. A vein valve typically consists of two:

ANSWER: cusps

34. Varicosities of veins in the anal region are called:

ANSWER: hemorrhoids

35. The three types of anastomoses that exist in the body are arterial, venous, and:

ANSWER: arteriovenous

36. Blood flow is directly proportional to a pressure difference and inversely proportional to:

ANSWER: resistance

37. Vascular resistance is proportional to the _____ of the blood.

ANSWER: viscosity

38. What is the name of the equation which combines the relationships of both pressure and resistance to blood flow?

ANSWER: Poiseuille-Hagen equation

39. The left ventricle pumps blood into the _____ circulation.

ANSWER: systemic

40. The right ventricle pumps blood into the _____ circulation.

ANSWER: pulmonary

41. Which vascular compartment has the greatest storage capacity, arterial or venous?

ANSWER: venous

42. The average blood pressure within the aorta is approximately _____ mmHg.

ANSWER: 100

43. The alternate expansion and recoil of an artery is called the:

ANSWER: arterial pulse

44. Locations on the body where applied pressure may stop arterial bleeding from a wound distal to that location are called:

ANSWER: pressure points

45. Clinically, blood pressure is measured with a:

ANSWER: sphygmomanometer

46. Tapping sounds that are the result of discontinuous turbulent blood flow are known clinically as the sounds of:

ANSWER: Korotkoff

47. Systemic blood pressure in the arteries at the peak of ventricular contraction is called the:

ANSWER: systolic pressure

48. The average systolic blood pressure in healthy adults at rest is:

ANSWER: 120 mmHg

49. At rest, the diastolic blood pressure of a healthy adult is approximately:

ANSWER: 70 to 80 mmHg

50. The difference between the systolic and diastolic blood pressures is called the:

ANSWER: pulse pressure

51. Diastolic pressure + 1/3 pulse pressure equals the:

ANSWER: mean arterial pressure

52. Cardiac output equals the product of heart rate and:

ANSWER: stroke volume

53. Resistance is inversely proportional to the fourth power of the vessel's:

ANSWER: radius

54. The vasomotor center is located in the:

ANSWER: medulla

55. The carotid sinus contains an abundant concentration of specialized nerve cells called:

ANSWER: baroreceptors (pressoreceptors)

56. Chemoreceptors consist of chemically sensitive cells clustered in tiny organs called carotid and:

ANSWER: aortic bodies

57. Kinins cause the smooth muscle cells of blood vessels to:

ANSWER: relax

58. The release of kinins results in a lowering of _____ resistance.

ANSWER: peripheral

59. A substance secreted by the atria, _____, causes a generalized vasodilation, which decreases peripheral resistance.

ANSWER: vasodilation

60. Hormone released by the adrenal medulla which has a generalized vasoconstrictor effect is:

ANSWER: Norepinephrine

61. Alcohol depresses the vasomotor center of the medulla, leading to decreased peripheral:

ANSWER: resistance

62. Alcohol inhibits the release of _____ from the posterior pituitary.

ANSWER: ADH

63. Angiotensin II is one of the most potent _____ known.

ANSWER: vasoconstrictors

64. The intermittent contraction and relaxation of precapillary sphincters is called:

ANSWER: vasomotion

65. Tissue _____ level is the principal factor regulating vasomotion.

ANSWER: oxygen

66. Forces operating in filtration-absorption dynamics are of two types: hydrostatic forces and _____ forces.

ANSWER: osmotic

TB.21:8

67. Which is greater, capillary hydrostatic pressure or interstitial fluid hydrostatic pressure?

ANSWER: capillary hydrostatic pressure

68. Plasma _____ contributes to a high osmotic pressure within the capillary.

ANSWER: protein

69. Which is lower, capillary osmotic pressure or interstitial fluid osmotic pressure?

ANSWER: interstitial fluid osmotic pressure

70. The type of shock in which the compensatory mechanisms are no longer successful is known as:

ANSWER: progressive shock

INSTRUCTIONS: ANSWER EACH QUESTION ACCORDING TO THE FOLLOWING KEY:

(A) Only 1 is correct
(B) Only 2 is correct
(C) Both are correct
(D) Neither are correct

71. The pulmonary trunk:

1. bifurcates into right and left pulmonary veins
2. carries relatively deoxygenated blood

ANSWER: B

72. Lobar arteries:

1. arise as branches of the pulmonary arteries
2. three serve the left lung

ANSWER: A

73. The pulmonary veins:

1. carry relatively deoxygenated blood
2. return blood to the right atrium

ANSWER: D

74. Superior vena cava:

    1. returns blood to right atrium
    2. carries venous blood from the upper
       limbs, head, and neck

ANSWER: C

75. Basilar artery:

    1. forms on ventral aspect of the brainstem
    2. arises through fusion of the two vertebral arteries

ANSWER: C

76. Tributary of the hepatic portal vein:

    1. superior mesenteric vein
    2. splenic vein

ANSWER: C

77. Direct branch of the abdominal aorta:

    1. subcostal artery
    2. posterior intercostal artery

ANSWER: D

78. Paired branch of the abdominal aorta:

    1. renal artery
    2. inferior mesenteric artery

ANSWER: A

79. Branch of the celiac artery:

    1. splenic artery
    2. left gastric artery

ANSWER: C

80. Branch of the celiac artery:

    1. common hepatic artery
    2. superior mesenteric artery

ANSWER: A

81. Branch of the common carotid artery:

    1. superior thyropid artery
    2. facial artery

ANSWER: D

82. Branch of the external carotid artery:

    1. maxillary artery
    2. posterior auricular artery

ANSWER: C

83. Axillary artery:

    1. begins at medial border of first rib
    2. becomes brachial artery at inferior border of teres major muscle

ANSWER: B

84. Brachial artery:

    1. arises as continuation of axillary artery
    2. gives rise to radial and ulnar arteries in forearm

ANSWER: C

85. Visceral branch of thoracic aorta:

    1. phrenic artery
    2. esophageal artery

ANSWER: C

86. Parietal branch of thoracic aorta:

    1. subcostal artery
    2. bronchial artery

ANSWER: A

87. Femoral artery:

    1. chief artery of the lower limb
    2. gives rise to popliteal artery

ANSWER: C

88. Medial plantar artery:

    1. arises directly from popliteal artery
    2. gives rise to dorsalis pedis artery

ANSWER: D

89. Anterior tibial artery:

    1. arises from popliteal artery
    2. gives rise to dorsalis pedis artery

ANSWER: C

90. Tributary of azygos vein:

    1. intercostal vein
    2. lumbar vein

ANSWER: C

91. Internal jugular vein:

    1. begins at the jugular foramen
    2. direct continuation of the sigmoid sinus

ANSWER: C

92. Found in the carotid sheath:

    1. vagus nerve
    2. external jugular vein

ANSWER: A

93. Median cubital vein:

    1. chief utilized for drawing blood samples
    2. tributary of basilic vein

ANSWER: C

94. Subclavian vein:

    1. begins at medial border of first rib
    2. direct continuation of the axillary vein

ANSWER: B

95. Brachiocephalic vein:

    1. formed through fusion of internal jugular and subclavian vein
    2. unites with contralateral partner to form the superior vena cava

ANSWER: C

96. Hepatic veins:

    1. carry blood to the liver from the digestive tract
    2. returns blood to the inferior vena cava

ANSWER: B

97. Inferior mesenteric vein:

    1. carries blood from the large intestine
    2. typically a tributary of the superior mesenteric vein

ANSWER: A

98. Hypotension:

    1. low arterial blood pressure
    2. often associated with circulatory shock

ANSWER: C

99. Atherosclerosis:

    1. most common form of arteriosclerosis
    2. may lead to cerebral vascular accidents

ANSWER: C

100. Angina pectoris:

    1. a symptom of cardiac ischemia
    2. pain often radiates to left shoulder and arm

ANSWER: C

101. Myocardial infarction:

    1. always the result of thrombus formation
    2. may lead to cardiac aneurysm

ANSWER: B

102. Cardiac tamponade:

    1. is not life-threatening
    2. accumulation of blood in pericardial cavity

ANSWER: B

103. Increased concentration in blood following a myocardial infarction:

    1. LDH
    2. SGOT

ANSWER: C

104. Derived from embryonic mesoderm:

    1. angioblasts
    2. blood vessels

ANSWER: C

105. Complication of atherosclerosis:

    1. aneurysm
    2. chronic (congestive) heart failure

ANSWER: C

TEST BANK for CHAPTER 22: THE LYMPHATIC SYSTEM

INSTRUCTIONS: ANSWER EACH QUESTION ACCORDING TO THE FOLLOWING KEY:

    (A) Only 1 is correct
    (B) Only 2 is correct
    (C) Both are correct
    (D) Neither are correct

1. The cisterna chyli:
    1. becomes thoracic duct at level of twelfth thoracic vertebra
    2. is a dilated lymph vessel within the thoracic region

ANSWER: A

2. Peyer's patches are found in the:
    1. distal ileum
    2. large intestine

ANSWER: A

3. The pharyngeal tonsil:
    1. is called an adenoid when enlarged
    2. is located at the base of the tongue

ANSWER: A

4. The palatine tonsils:
    1. possess numerous pits called crypts
    2. are termed adenoids when enlarged

ANSWER: A

5. The functions of lymph nodes include:
    1. production of lymphocytes
    2. filtering of lymph

ANSWER: C

6. The cortex of a typical lymph node:

    1. contains germinal centers
    2. contains lymph sinuses

ANSWER: C

7. The trabeculae of lymph nodes:

    1. provides support for blood vessels
    2. are composed of connective tissue

ANSWER: C

8. The deep cervical lymph nodes:

    1. drain only a portion of the lymph vessels of the head and neck
    2. are located primarily on the deep surface of the trapezius muscle

ANSWER: D

9. Major groups of lymph nodes responsible for draining the body wall include:

    1. lumbar lym[ph nodes
    2. common iliac nodes

ANSWER: C

10. The axillary lymph nodes drain lymph vessels from the:

    1. breast
    2. upper limb

ANSWER: C

11. The spleen:

    1. contains macrophages
    2. functions in part as a reservoir for the blood

ANSWER: C

12. Hassall's corpuscles:

    1. are located in the spleen
    2. are small groups of epithelial cells

ANSWER: B

13. The thymus:

    1. consists of two lobes
    2. begins to involute at puberty

ANSWER: C

14. The spleen:

    1. is not vital to life
    2. may be the source of a great blood loss should it be ruptured

ANSWER: C

15. The thymus:

    1. consists of an outer cortex and an inner medulla
    2. receives stem cells from the bone marrow and liver prior to birth

ANSWER: C

16. Lymph is formed:

    1. from interstitial fluid
    2. directly from plasma

ANSWER: A

17. The disease called filariasis:

    1. is often referred to as elephantiasis
    2. is caused by larval nematodes that are transmitted to humans by misquitoes

ANSWER: C

18. The thymus is:

    1. larger in adulthood than in the newborn
    2. located in the superior mediastinum

ANSWER: B

19. Splenic white pulp:

    1. consists primarily of reticular fibers
    2. contains dense masses of lymphocytes

ANSWER: B

20. The stroma of the spleen consists of the:

    1. trabeculae
    2. connective tissue capsule

INSTRUCTIONS: TRUE - FALSE

21. (T/F): The largest organ of the lymphatic system is called the spleen.

ANSWER: T

22. (T/F): Lymph nodes filter lymph and produce lymphocytes.

ANSWER: T

23. (T/F): Lymph nodes are often times referred to as lymph glands.

ANSWER: T

24. (T/F): The circulation of lymph through a node involves three types of lymph vessels.

ANSWER: T

25. (T/F): Afferent lymph vessels convey lymph away from a node.

ANSWER: F

26. (T/F): Efferent lymph vessels do not possess valves.

ANSWER: F

27. (T/F): Lymph sinuses interconnect the afferent and efferent lymph vessels of a node.

ANSWER: T

28. (T/F): Infection may result in enlarged and tender lymph nodes.

ANSWER: T

29. (T/F): Retroauricular nodes are located near the mastoid process of the temporal bone.

ANSWER: T

30. (T/F): Germinal centers are not found to be associated with lymph nodes.

ANSWER: F

31. (T/F): Afferent lymph vessels typically enter a node through its hilus.

ANSWER: F

32. (T/F): Lymph flows from a node by way of the efferent lymph vessel.

ANSWER: T

33. (T/F): Lymph nodes typically occur in clusters or chains along lymphatics.

ANSWER: T

34. (T/F): Cancer cells are often transported from a primary tumor to other regions of the body via lymphatics.

ANSWER: T

35. (T/F): Popliteal lymph nodes are situated on the anterior aspect of the knee.

ANSWER: F

36. (T/F): The inguinal group of lymph nodes consists of a deep and a superficial group.

ANSWER: T

37. (T/F): "Blood poisoning" is more appropriately called lymphangitis.

ANSWER: T

38. (T/F): Blood leaving the spleen in the splenic vein is delivered to the hepatic portal vein.

ANSWER: T

39. (T/F): Any obstruction of the lymphatic vessels can lead to a condition called edema.

ANSWER: T

40. (T/F): The principal force moving the plasma out of the blood is the osmotic pressure.

ANSWER: F

COMPLETE THE FOLLOWING STATEMENTS

41. The portion of the plasma the escapes through the capillary walls and enters the tissues is called:

ANSWER: interstitial fluid

42. Interstitial fluid is produced from the blood:

ANSWER: plasma

43. Most of the fluid in the body is located within the cells themselves, the intracellular:

ANSWER: fluid (compartment)

44. The thymus instructs the T lymphocyte to distinguish between 'self' and:

ANSWER: 'nonself'

45. Surgical removal of the spleen is called:

ANSWER: splenectomy

46. Surgical removal of the thymus is called:

ANSWER: thymectomy

47. Germinal centers contain immature:

ANSWER: lymphocytes

48. Aggregates of lymph nodules located at the base of the tongue are referred to as the:

ANSWER: lingual tonsils

49. Lymph nodules located in the distal ileum are called:

ANSWER: Peyer's patches

50. Lymph tissue is arranged in small aggregates called nodes and:

ANSWER: nodules

51. The outer connective tissue covering of a lymph node is designated as its:

ANSWER: capsule

52. At the level of the twelfth thoracic vertebra, the cisterna chyli narrows and becomes the:

ANSWER: thoracic duct

53. The thoracic duct typically delivers lymph to the junction of the left subclavian vein and the left:

ANSWER: internal jugular vein

54. Utilization of radiopaque material to visualize lymphatic vessels is a proceedure called:

ANSWER: lymphangiography

55. Each lymph capillary originates as a blind or closed:

ANSWER: tube

56. Large lymphatics that drain groups of lymph nodes are called:

ANSWER: lymphatic trunks

57. Lymph capillaries unite with one another to form larger vessels called lymphatics, or:

ANSWER: lymph veins

58. The arterial supply to a lymph node enters the organ through its:

ANSWER: hilus

59. The supratrochlear lymph nodes are located in the region of the:

ANSWER: elbow

60. Hassall's corpuscles are small groups of epithelial cells located in the splenic:

ANSWER: medulla

INSTRUCTIONS: SELECT THE CORRECT RESPONSE FROM THE COLUMN ON THE RIGHT. A RESPONSE MAY BE USED ONCE, MORE THAN ONCE OR NOT AT ALL

61.-(b)- a hormone

62.-(d)- cells located in thymic medulla

63.-(f)- tissue fluid

64.-(c)- largest lymphatic organ

65.-(a)- edema

a. swelling

b. thymosin

c. spleen

d. Hassall's corpuscles

e. lymph

f. interstitial fluid

g. thymus

TEST BANK for CHAPTER 23: THE BODY'S DEFENSE MECHANISMS:IMMUNITY

INSTRUCTIONS: TRUE – FALSE

1. (T/F): Defense mechanisms of the body can be nonspecific or highly specific.

ANSWER: T

2. (T/F): A substance capable of stimulating an immune response is called an antigen.

ANSWER: T

3. (T/F): Nonspecific mechanisms of defense include interferon release, inflammation, and phagocytosis

ANSWER: T

4. (T/F): Natural killer cells are not stimulated by the release of interferons.

ANSWER: F

5. (T/F): Natural killer cells recognize cells of the body that have been altered by viruses and destroy them.

ANSWER: T

6. (T/F): The clinical characteristics of inflammation include redness, heat, edema and pain.

ANSWER: T

7. (T/F): Interleukin 1 is a steroid hormone released by all lymphocytes.

ANSWER: F

8. (T/F): Fever increases viral activity.

ANSWER: F

9. (T/F): Fever increases the level of circulating iron within the body.

ANSWER: F

10. (T/F): The membrane bound vesicle within a marcophage that contains an injested bacterium is called a phagosome.

ANSWER: T

11. (T/F): Fusion of lysosomes with a phagosome membrane is a process called degranulation.

ANSWER: T

12. (T/F): Cell-mediated immunity is one form of nonspecific defense.

ANSWER: F

13. (T/F): Antibody-mediated immunity is one of the body's specific defense mechanisms.

ANSWER: T

14. (T/F): Memory B cells continue to produce small quantities of antibody for years after and infection.

ANSWER: T

15. (T/F): Interleukin 2 is secreted by helper T cells.

ANSWER: T

16. (T/F): Natural killer cells are responsible for the secretion of B-cell growth factor.

ANSWER: F

17. (T/F): Included among the large granular lymphocytes are cells known as natural killer or NK cells.

ANSWER: T

18. (T/F): Suppressor T cells inhibit the activity of
    natural killer cells.

ANSWER: F

19. (T/F): Cytotoxic T cells are inhibited by suppressor T cells.

ANSWER: T

20. (T/F): Plasma cells are the cells responsible for the
    production of antibody.

ANSWER: T

21. (T/F): It is easy to distinguish between a T and B cell with
    a ordinary light microscope.

ANSWER: F

22. (T/F): Functional B cells are lacking in the clinical condition
    called DiGeorge's syndrome.

ANSWER: F

23. (T/F): Deficient B cell function is characteristic of the
    clinical syndrome called Burton's agammaglobulinemia.

ANSWER: T

24. (T/F): Interleukin 1 plays a role in the production of fever.

ANSWER: T

25. (T/F): Human leukocyte antigen markers are present on the
    surface of all the body's cells.

ANSWER: T

26. (T/F): Antigen-presenting cells are macrophages.

ANSWER: T

27. (T/F): Antibodies are highly specific proteins called
    immunoglobulins.

ANSWER: T

28. (T/F): A typical immunoglobulin consists of six polypeptide chains.

ANSWER: F

29. (T/F): Haptens are substance that become antigenic when they attach to proteins.

ANSWER: T

30. (T/F): There are ten classes of antibodies.

ANSWER: F

31. (T/F): In humans approximately 75% of the antibodies in the blood belong to the IgM class.

ANSWER: F

32. (T/F): The principal antibody found in body secretions is of the IgG class.

ANSWER: F

33. (T/F): The C region of the five major antibody classes determines their functional activity.

ANSWER: T

34. (T/F): A given antigen may bind with different affinities to different antibodies.

ANSWER: T

35. (T/F): A typical antibody has a circular configuration.

ANSWER: F

36. (T/F): The abbreviation Ig stands for immunoglobulin.

ANSWER: T

37. (T/F): Antibodies of the IgE group are mediators of the allergic response.

ANSWER: T

38. (T/F): Cells responsible for cell-mediated immunity include the T cells and monocyte/macrophage system.

ANSWER: T

39. (T/F): Patients with ARC may eventually develop AIDS, but the extent of this risk is not yet known.

ANSWER: T

40. (T/F): Acquired immune deficiency syndrome results from an infection with a retrovirus.

ANSWER: T

41. (T/F): A retrovirus use its own RNA as a template to make DNA with the help of an enzyme called reverse transcriptase.

ANSWER: T

42. (T/F): Approximately 90% of patients with AIDS develop a neuropsychological disorder called AIDS dementia.

ANSWER: T

43. (T/F): Transmission of AIDS may occur through direct exposure to infected blood or blood products.

ANSWER: T

44. (T/F): The AIDS virus rapidly infects suppressor T cells.

ANSWER: F

45. (T/F): Acquired immune deficiency syndrome is not typically spread by casual contact.

ANSWER: T

46. (T/F): Acquired immune deficiency syndrome may be transmitted by semen during sexual intercourse.

ANSWER: T

47. (T/F): Perforins are cytotoxic proteins that are secreted by cytotoxic T cells.

ANSWER: T

48. (T/F): Lymphotoxins are believed to be especially toxic for cancer cells.

ANSWER: T

49. (T/F): Active immunity can be induced by the process of immunization.

ANSWER: T

50. (T/F): Active immunity is only induced artificially through immunization.

ANSWER: F

INSTRUCTIONS: ANSWER EACH QUESTION ACCORDING TO THE FOLLOWING KEY:

(A) Only 1 is correct
(B) Only 2 is correct
(C) Both are correct
(D) Neither are correct

51. Active immunity may be induced:

1. artificially
2. naturally

ANSWER: C

52. Blocking antibodies:

1. are produced by natural killer cells
2. prevent T cells from recognizing certain cancer cells

ANSWER: B

53. Tissues involved in an autograft are:

1. compatible
2. of the same HLA type

ANSWER: C

54. Cyclosporin A:

    1. is an antibiotic extracted from fungi
    2. inhibits B cell function markedly

ANSWER: A

55. Immunologically privileged location:

    1. brain
    2. cornea

ANSWER: C

56. Autoimmune disease:

    1. Grave's disease
    2. Systemic lupus erythematosus

ANSWER: C

57. Allergic reaction:

    1. systemic anaphylaxis
    2. hives

ANSWER: C

58. Antihistamines:

    1. compete for histamine receptor sites on cells
    2. are compeletly effective when used clinically

ANSWER: A

59. Reagins are:

    1. specific immunoglobulins of the IgE class
    2. secreted by helper T cells

ANSWER: A

60. The secondary immune response:

    1. may be elicited years after the first exposure to an antigen
    2. is typically more rapid to develop than the primary response

ANSWER: C

COMPLETE THE FOLLOWING STATEMENTS

61. The first exposure to an antigen typically elicits a:

ANSWER: primary immune response

62. A virus that is weakened by successive passages through cells of nopnhuman hosts is said to be:

ANSWER: attenuated

63. An individual given antibodies actively produced by another organism is said to have:

ANSWER: passive immunity

64. The complement system may be activated in the absence of antigen-antibody complexes by way of the:

ANSWER: properdin pathway

65. Some proteins of the complement system coat pathogens in a process called:

ANSWER: opsinization

66. Each polypeptide chain of the immunoglobulin molecule has a constant region, junctional region, and a:

ANSWER: variable region

67. The constant region of an immunoglobulin molecule anchors the molecule to the:

ANSWER: plasma membrane

68. Antigens that have the same determinant repeated several times are said to be:

ANSWER: multivalent

69. Lymphocytes that activly produce and secrete antibody are called:

ANSWER: plasma cells

70. The rapidity of a secondary immune response is due in part to the persistence of:

ANSWER: memory B cells

INSTRUCTIONS: ANSWER EACH QUESTION ACCORDING TO THE FOLLOWING KEY:

(A) Only 1 is correct
(B) Only 2 is correct
(C) Both are correct
(D) Neither are correct

71. Nonspecific defense mechanism:

1. inflammation
2. stomach enzymes

ANSWER: C

72. Nonspecific defense mechanism:

1. acid secretion by the stomach
2. antibody production by plasma cells

ANSWER: A

73. Nonspecific defense mechanism:

1. microorganisms of the skin
2. interferon release

ANSWER: C

74. Clinical sign of inflammation:

    1. pain
    2. edema

ANSWER: C

75. Cytotoxic T cells:

    1. also referred to as killer T cells
    2. inhibit activity of helper T cells

ANSWER: A

76. Immunoglobulin light chain:

    1. composed of approximately 400 amino acids
    2. has a constant region, variable region, and junctional region

ANSWER: B

77. Heavy and light immunoglobulin chains are held together by:

    1. noncovalent bonds
    2. disulfide linkages

ANSWER: C

78. The immunoglobulin constant, or C region:

    1. contains the site for antigen binding
    2. has the same amino acid sequence from one immunoglobulin type to another

ANSWER: B

79. Memory B cells:

   1. activated B cells that do not differentiate into plasma cells
   2. important for the secondary immune response

ANSWER: C

80. Class I antigens:

   1. found on all nucleated cells of the body
   2. more limited in distribution than Class II antigens

ANSWER: A

TEST BANK for CHAPTER 24: THE RESPIRATORY SYSTEM

INSTRUCTIONS: ANSWER EACH QUESTION ACCORDING TO THE FOLLOWING KEY:

   (A) Only 1 is correct
   (B) Only 2 is correct
   (C) Both are correct
   (D) Neither are correct

1. Contribute to the external shape of the nose:

   1. alar cartilages
   2. nasal conchae

ANSWER: A

2. One of the stages of respiration:

   1. pulmonary ventilation
   2. cellular respiration

ANSWER: C

3. Component of the upper respiratory tract:

   1. pharynx
   2. trachea

ANSWER: A

4. Component of the lower respiratory tract:

   1. pharynx
   2. larynx

ANSWER: B

5. Forms part of the upper respiratory tract:

   1. nasal cavity
   2. pharynx

ANSWER: C

6. The nostrils:

    1. are also termed the external nares
    2. lead to a cavity called the vestibule

ANSWER: C

7. Component of the nasal septum:

    1. vomer
    2. nasal conchae

ANSWER: A

8. Forms portion of the nasal septum:

    1. maxillae
    2. perpendicular plate of the ethmoid

ANSWER: C

9. The nasal septum is formed partially by:

    1. cartilage
    2. the vomer

ANSWER: C

10. Bone containing paranasal sinus:

    1. ethmoid
    2. zygomatic (malar)

ANSWER: A

11. Bone containing paranasal sinus:

    1. sphenoid
    2. maxilla

ANSWER: C

12. Inflammation of the mucous membrane of the nose:

    1. is termed sinusitis
    2. may be caused by infection or irritants

ANSWER: B

13. The nasal mucous membrane:

    1. is pseudostratified ciliated columnar epithelium
    2. contains goblet cells

ANSWER: C

14. The floor of the nasal cavity is formed by the:

    1. hard palate
    2. soft palate

ANSWER: C

15. The oropharynx functions in:

    1. digestion
    2. respiration

ANSWER: C

16. The nasopharynx:

    1. functions in respiration
    2. is lined by pseudostratified ciliated columnar epithelium

ANSWER: C

17. The nasopharynx contains the:

    1. openings for the eustachian tubes
    2. pharyngeal tonsils

ANSWER: C

18. The oropharynx:

    1. is lined by pseudostratified ciliated columnar epithelium
    2. contain the pharyngeal tonsils

ANSWER: D

19. The oropharynx contains the:

    1. palatine tonsils
    2. lingual tonsils

ANSWER: C

20. The most inferior portion of the pharynx is:

    1. lined by stratified squamous epithelium
    2. called the laryngopharynx

ANSWER: C

21. The larynx is:

    1. also termed the 'voicebox'
    2. consists in part of nine cartilages

ANSWER: C

22. Regarding the epiglottis:

    1. it is leaf-shaped
    2. it is composed of cartilage

ANSWER: C

23. Regarding the glottis:

    1. covered by epiglottis during swallowing
    2. food entering it may end up in the trachea

ANSWER: C

24. The thyroid cartilage:

    1. is the smallest of the laryngeal cartilages
    2. is located superior to the cricoid cartilage

ANSWER: B

25. Paired laryngeal cartilages:

    1. arytenoid
    2. cricoid

ANSWER: A

26. Unpaired laryngeal cartilage:

    1. corniculate
    2. cuneiform

ANSWER: D

27. The true vocal cords:

    1. are called the ventricular folds
    2. surround the opening called the rima glottidis

ANSWER: B

28. The ventricular folds:

    1. are located superior to the true vocal cords
    2. keep the larynx closed during swallowing

ANSWER: C

29. Structure helping to modify the sounds of speech:

    1. paranasal sinuses
    2. muscles of the cheeks

ANSWER: C

30. Regarding the trachea:

    1. unites larynx with primary bronchi
    2. approximately 22 cm in length

ANSWER: A

31. Regarding the trachea:

    1. lies posterior to the esophagus
    2. lined by pseudostratified ciliated columnar epithelium

ANSWER: B

32. The mucous membrane lining of the trachea is:

    1. often called the mucociliary escalator
    2. known to contain goblet cells

ANSWER: C

TB.24:6

33. Regarding the left primary bronchus:

    1. shorter than the right primary bronchus
    2. wider than the right primary bronchus

ANSWER: D

34. Regarding the primary bronchi:

    1. both are considered to be extrapulmonary in their location
    2. give rise to secondary bronchi

ANSWER: C

35. Regarding the trachea:

    1. its most distal cartilage is called the carina
    2. divides at the sternal angle to form the primary bronchi

ANSWER: C

36. The tracheal cartilages:

    1. are open in a posterior direction
    2. consist largely of fibrocartilage

ANSWER: A

37. Tertiary bronchi:

    1. give rise to bronchioles
    2. arise from secondary bronchi

ANSWER: C

38. Regarding the lungs:

    1. the cardiac notch is a characteristic feature of the right lung
    2. the costal surface is associated with the ribs

ANSWER: B

39. Each lung has:

    1. an oblique fissure
    2. three lobes

ANSWER: A

40. Regarding the lungs:

    1. the apex is directed superiorly
    2. separated from one another by the mediastinum

ANSWER: C

41. The parietal pleura:

    1. is continuous with the visceral pleura
    2. is in direct contact with the surface of the lung

ANSWER: A

42. Regarding the pleura:

    1. visceral pleura is associated with the surface of the lungs
    2. enclose a potential space called the pleural cavity

ANSWER: C

43. Inflammation of the pleura:

    1. is called pleurisy
    2. may be a complication of another pulmonary disease

ANSWER: C

44. A component of the repiratory membrane:

    1. alveolar membrane
    2. pulmonary capillary basement membrane

ANSWER: C

45. Regarding the condition known as emphysema:

    1. lungs become more elastic than usual
    2. scar tissue and air pockets form in the lungs

ANSWER: B

46. Movement of air into and out of the lungs:

   1. ventilation
   2. expiration

ANSWER: A

47. Intrapleural pressure:

   1. approximately 770 mmHg
   2. normally below atmospheric pressure

ANSWER: B

48. Regarding Boyle's law:

   1. pressure of gas varies inversely with volume of gas with temperature being constant
   2. as the volume increases the pressure decreases

ANSWER: C

49. Regarding inspiration:

   1. intrapleural pressure increases
   2. diaphragm relaxes and moves inferiorly

ANSWER: D

50. Regarding expiration:

   1. diaphragm contracts and moves superiorly
   2. thorcic volume decreases

ANSWER: B

INSTRUCTIONS: TRUE - FALSE

51. (T/F): The nasal bones form the bridge of the nose.

ANSWER: T

52. (T/F): Pulmonary surfactant is produced by septal cells of the alveolar epithelium.

ANSWER: T

53. (T/F): A low compliance means that the lungs and thorax are easily expandible.

ANSWER: T

54. (T/F): The sum of the tidal volume and the expiratory reserve volume is the vital capacity.

ANSWER: F

55. (T/F): Inability to breath in a horizontal position is known clinically as eupnea.

ANSWER: F

56. (T/F): In the healthy individual the physiological and anatomical dead spaces are approximately equal in volume.

ANSWER: T

57. (T/F): The partial pressure of carbon dioxide in the atmosphere, according to Dalton's law, is greater than that of oxygen.

ANSWER: F

58. (T/F): Breathing rate per minute times tidal volume equals the minute respiratory volume (MRV).

ANSWER: T

59. (T/F): Each hemoglobin molecule contains four heme groups.

ANSWER: T

60: (T/F): A decrease in blood pH will shift the oxyhemoglobin dissociation curve to the left.

ANSWER: F

61. (T/F): An increase in blood pH will shift the oxyhemoglobin curve to the left.

ANSWER: T

62. (T/F): Hypoxic hypoxia may result from an obstructed airway, breathing at high altitudes, or the presence of fluid in the lungs.

ANSWER: T

63. (T/F): Fetal hemoglobin has a markedly lower affinity for oxygen than does adult hemoglobin.

ANSWER: F

64. (T/F): The oxyhemoglobin dissociation curve is not affected by temperature.

ANSWER: T

65. (T/F): Reduced hemoglobin has a higher affinity for carbon dioxide than does fully saturated oxyhemoglobin.

ANSWER: T

INSTRUCTIONS: COMPLETE THE FOLLOWING:

66. That part of the brainstem containing the pneumotaxic and apneustic centers is called the:

ANSWER: pons

67. That part of the brainstem containing the respiratory rhythmicity areas is called the:

ANSWER: medulla

68. An important protective mechanism involving stretch receptors and preventing overfilling of the lungs is known as the:

ANSWER: Hering-Breuer reflex

69. Tertiary bronchi supply an area of the lung called a:

ANSWER: bronchopulmonary segment

70. The terminal bronchioles divide within lobules to form:

ANSWER: respiratory bronchioles

71. Respiratory bronchioles give rise to the next division of the respiratory tree called the:

ANSWER: alveolar ducts

72. A substance, produced by alveolar septal cells, that helps to reduce surface tension is called:

ANSWER: surfactant

73. The sum of the tidal volume and the expiratory reserve volume is called the:

ANSWER: expiratory capacity

74. The largest amount of air that can be ventilated is called the:

ANSWER: vital capacity

75. The volume of air that can be inspired above and beyond the normal tidal volume is called the:

ANSWER: inspiratory reserve volume

76. The anatomical dead space normally has a volume of about:

ANSWER: 150 milliliters

77. Hemoglobin bound to oxygen is called:

ANSWER: oxyhemoglobin

78. Surgical opening of the trachea and insertion of a tube to facilitate breathing is known as a:

ANSWER: tracheostomy

79. Respiratory distress syndrome of the newborn is also called:

ANSWER: hyaline membrane disease

80. Bluish discoloration of the skin and mucous membranes due to high amounts of reduced hemoglobin is called:

ANSWER: cyanosis

INSTRUCTIONS: SELECT THE SINGLE BEST ANSWER

81. What percentage of carbon dioxide is normally transported attached to hemoglobin?

   a) 10
   b) 20
   c) 35
   d) 45

ANSWER: b

82. The inspiratory capacity of the normal lungs is approximately:

   a) 1500 milliliters
   b) 2200 milliliters
   c) 3500 milliliters
   d) 4500 milliliters

ANSWER: c

83. The vital capacity of normal lungs is approximately:

   a) 2700 milliliters
   b) 3700 milliliters
   c) 4700 milliliters
   d) 5700 milliliters

ANSWER: c

84. The volume of the functional residual capacity is typically about:

   a) 2400 milliliters
   b) 3400 milliliters
   c) 4400 milliliters
   d) 5400 milliliters

ANSWER: a

85. During normal inspiration, a person inspires about what volume of air?

   a) 50 milliliters
   b) 100 milliliters
   c) 250 milliliters
   d) 500 milliliters

ANSWER: d

86. To what extent does carbonic anhydrase catalyze the reaction in which carbon dioxide combines with water to form carbonic acid?

   a) 200 fold
   b) 500 fold
   c) 1000 fold
   d) 5000 fold

ANSWER: d

87. With increased metabolic activity of a given cell:

   a) the partial pressure of carbon dioxide within
      the cell increases
   b) the pH within the cell increases
   c) the temperature within the cell declines
   d) the cell requires less oxygen

ANSWER: a

88. The partial pressure of carbon dioxide in alveolar air is typically:

   a) 5 mmHg
   b) 10 mmHg
   c) 20 mmHg
   d) 40 mmHg

ANSWER: d

89. The partial pressure of oxygen in inspired air is typically:

   a) 50 mmHg
   b) 100 mmHg
   c) 160 mmHg
   d) 260 mmHg

ANSWER: c

90. The partial pressure of oxygen in expired air is typically:

   a) 50 mmHg
   b) 120 mmHg
   c) 200 mmHg
   d) 250 mmHg

ANSWER: b

TB.25:1

TEST BANK for CHAPTER 25: THE DIGESTIVE SYSTEM

INSTRUCTIONS: ANSWER EACH QUESTION ACCORDING TO THE FOLLOWING KEY:

    (A) Only 1 is correct
    (B) Only 2 is correct
    (C) Both are correct
    (D) Neither are correct

1. Regarding the myenteric plexus of Auerbach:

    1. located within the submucosal layer of the digestive tract
    2. an autonomic nerve plexus

ANSWER: B

2. The visceral peritoneum:

    1. is a serous membrane
    2. is continuous with the parietal peritoneum

ANSWER: C

3. The clinical condition known as ascites:

    1. is also called peritonitis
    2. is characterized by the accumulation of fluid within the peritoneal cavity

ANSWER: B

4. The oral cavity:

    1. is the space between the teeth and the lips or cheeks
    2. is also called the buccal cavity

ANSWER: B

5. The permanent dentition:

    1. replaces the deciduous dentition
    2. consists of 32 teeth

ANSWER: C

6. The deciduous dentition:

   1. begins to emerge at approximately six months of age
   2. consists of 20 teeth

ANSWER: C

7. The process of mastication:

   1. is the act of chewing food
   2. serves as an aid to digestion by increasing the surface area of the food

ANSWER: C

8. Zymogenic (Chief) cells of the gastric mucosa:

   1. secrete hydrochloric acid
   2. produce a substance called gastric lipase

ANSWER: B

9. Contain taste buds:

   1. fungiform papillae
   2. filiform papillae

ANSWER: A

10. A tooth consists mainly of hard, calcified:

    1. dentin
    2. enamel

ANSWER: A

11. The dentin of the root of a tooth is covered by a bonelike substance called:

    1. pulp
    2. cementum

ANSWER: B

12. Contained within the pulp cavity of a tooth:

    1. lymphatics
    2. nerves

ANSWER: C

13. In terms of their dentition, humans are classified as:

    1. diphyodonts
    2. heterodonts

ANSWER: C

14. Type of tooth forming part of the permanent dentition:

    1. canine
    2. incisor

ANSWER: C

15. Each quadrant of the adult permanent dentition contains how many teeth?

    1. seven
    2. eight

ANSWER: B

16. Secrete a substance called saliva:

    1. buccal glands
    2. parotid glands

ANSWER: C

17. Ducts of the submandibular glands:

    1. are called Rivinus's ducts
    2. are located near the base of the tongue

ANSWER: B

18. The sublingual glands:

    1. are the largest of the salivary glands
    2. have ducts called Wharton's ducts

ANSWER: D

19. Salivary amylase:

    1. initiates the digestion of carbohydrates
    2. splits starch molecules into fragments called dextrins

ANSWER: C

20. The parotid gland:

    1. secretes an enzyme called salivary amylase
    2. opens into the oral cavity via Stensen's duct

ANSWER: C

21. Salivary amylase:

    1. has an optimum pH of 8.0
    2. is the primary secretory product of the sublingual glands

ANSWER: D

22. The parotid duct:

    1. opens into the oral cavity opposite the upper second molar tooth
    2. is also called Wharton's duct

ANSWER: A

23. The secretion of saliva:

    1. is a reflex response
    2. involves parasympathetic neurons located within the medulla and pons

ANSWER: C

24. Cells of the salivatory nuclei:

    1. increase the rate of saliva secretion
    2. are predominantly sympathetic neurons

ANSWER: A

25. The release of saliva can be initiated by:

    1. visual stimuli
    2. olfactory stimuli

ANSWER: C

INSTRUCTIONS: TRUE - FALSE

26. (T/F): Hydrolysis is the breaking down of a substance into smaller units through the addition of water.

ANSWER: T

27. (T/F): Fructose is the major monosaccharide utilized by the body.

ANSWER: F

28. (T/F): Lipases are enzymes utilized in the digestion of fats.

ANSWER: T

29. (T/F): Auditory stimuli can elicit the release of saliva.

ANSWER: T

30. (T/F): The reflex release of saliva may only be initiated through the stimulation of taste receptors.

ANSWER: F

31. (T/F): The laryngopharynx represents the superior aspect of a muscular tube called the esophagus.

ANSWER: F

32. (T/F): The esophagus is composed entirely of smooth muscle fibers.

ANSWER: F

33. (T/F): The esophagus is a muscular tube of approximately twenty inches in length.

ANSWER: F

34. (T/F): The esophagus interconnects the pharynx and the stomach.

ANSWER: T

35. (T/F): The primary function of the esophagus is to mechanically mix the injested food bolus.

ANSWER: F

36. (T/F): The last stage of deglutition is called the esophageal stage.

ANSWER: T

37. (T/F): The esophageal stage is the shortest step in the three-stage process of deglutition.

ANSWER: F

38. (T/F): The esophagus exhibits peristaltic contractions during the deglution process.

ANSWER: T

39. (T/F): The lateral border of the stomach is also called the greater curvature.

ANSWER: T

40. (T/F): Food enters the stomach through its cardiac end.

ANSWER: T

41. (T/F): The portion of the stomach extending above the level of the cardiac orifice is designated as the antrum.

ANSWER: F

42. (T/F): Food typically remians in the stomach for about ten to fifteen hours prior to entering the duodenum.

ANSWER: F

43. (T/F): In an empty stomach the mucosa and submucosa exist as longitudinal folds called the rugae.

ANSWER: T

44. (T/F): The thorough churning of food by the contracting stomach leads to the formation of a semiliquid mixture called chyme.

ANSWER: T

45. (T/F): Enteroendocrine cells are found in gastric glands.

ANSWER: T

46. (T/F): Parietal cells of the gastric glands are responsible for the secretion of hydrochloric acid.

ANSWER: T

47. (T/F): Parietal cells of the gastric glands are also called oxyntic cells.

ANSWER: T

48. (T/F): The mechanical breakdown of food by the teeth results in the formation of a semiliquid mixture called chyme.

ANSWER: F

49. (T/F): Pepsin has an optimum pH of approximately 6.5

ANSWER: F

50. (T/F): The major digestive activity occurring in the stomach is the breakdown of fats.

ANSWER: F

51. (T/F): The only proteolytic enzyme of the stomach that is capable of digesting collagen is pepsin.

ANSWER: T

52. (T/F): Gastric lipase has a pH optimum of 2 to 3.

ANSWER: F

53. (T/F): Mucus-secreting cells play a role in protecting the lining of the stomach from the acidic environment.

ANSWER: T

54. (T/F): Gastrin release can be stimulated by the presence of alcohol in the stomach.

ANSWER: T

55. (T/F): Gastric motility may be increased by the secretion of gastrin.

ANSWER: T

56. (T/F): The amount of hydrochloric acid secreted during a meal is directly proportional to the amount of protein in the meal.

ANSWER: T

57. (T/F): Protein digestion products stimulate the release of gastrin which also causes an increase in HCl secretion.

ANSWER: T

58. (T/F): The entergastric reflex is responsible for the stimulation of gastric secretion and motility.

ANSWER: F

59. (T/F): Gastrin facilitates gastric emptying.

ANSWER: T

60. (T/F): Secretin and cholecystokinin are intestinal hormones.

ANSWER: T

INSTRUCTIONS: COMPLETE THE FOLLOWING:

61. The presence of chyme in the stomach initiates a reflex response called the:

ANSWER: enterogastric reflex

62. The tail of the pancreas is directed to the left near the hilum of the:

ANSWER: spleen

63. The exocrine cells of the pancreas account for what percentage of the pancreatic cells?

ANSWER: 99%

64. The accessory pancreatic duct is also called the duct of:

ANSWER: Santorini

65. The pancreatic duct unites with the common bile duct to form a structure called the:

ANSWER: hepatopancreatic ampulla

66. Blood-filled spaces located between the plates of hepatic cells are known as:

ANSWER: sinusoids

67. Phagocytic cells associated with the hepatic sinusoids are called:

ANSWER: Kupffer cells

68. The right and left hepatic ducts unite to form a structure called the:

ANSWER: common hepatic duct

69. The common hepatic duct joins the cystic duct to form a structure called the:

ANSWER: common bile duct

70. The common bile duct is formed as the common hepatic duct joins with the:

ANSWER: cystic duct

71. Chyme and digestive solutions are thoroughly mixed in the small intestine by a contractile process called:

ANSWER: segmentation

72. The shortest portion of the small intestine is called the:

ANSWER: doudenum

73. The first (proximal) portion of the small intestine is called the:

ANSWER: duodenum

74. The portion of the small intestine which joins the colon is called the:

ANSWER: ileum

75. Within the submucosa of the duodenum are mucous glands called:

ANSWER: Brunner's glands

76. The substance in pancreatic juice which serves to neutralize the acidic chyme in the duodenum is called:

ANSWER: sodium bicarbonate

77. Tyrpsinogen is converted to trypsin by an enzyme produced by the intestinal mucosa called:

ANSWER: enterokinase

78. The process of breaking down large fat droplets into many smaller fat droplets is called:

ANSWER: emulsification

79. Cholecystokinin results in the relaxation of the sphincter of:

ANSWER: Oddi

80. Maltose is broken down into two glucose molecules by the enzyme,:

ANSWER: maltase

81. The major enzyme responsible for the digestion of fats is called:

ANSWER: pancreatic lipase

82. Nucleic acids are digested by ribonuclease and deoxyribonuclease into pentoses and:

ANSWER: nitrogenous bases

83. Glucose and galactose are actively absorbed while fructose is absorbed by:

ANSWER: facilitated diffusion

84. Triglycerides combine with cholesterol, lipoprotein, and phospholipids to form globules called:

ANSWER: chylomicrons

85. The fat-soluble vitamins include vitamins A, D, K and:

ANSWER: E

86. The sigmoid colon has its own mesentery, the:

ANSWER: sigmoid mesocolon

87. The anal sphincter that is under voluntary control is the:

ANSWER: external anal sphincter

88. The major bile pigment excreted by the large intestine is called:

ANSWER: bilirubin

89. Inflammation of the liver due to viruses, drugs, and chemicals is clinically known as:

ANSWER: hepatitis

90. A clinical condition characterized by uncontrollable overeating usually followed by forced vomiting is known as:

ANSWER: bulimia

91. The primitive gut in the developing embryo is lined by which germ layer?

ANSWER: endoderm

92. During embryonic development the proctodeum becomes the:

ANSWER: anus

93. The clinical term for an inability to swallo is:

ANSWER: aphagia

94. The mucous membrane of the anal canal is arranged in longitudinal folds called the:

ANSWER: anal columns

95. The ascending colon joins the transverse colon at the:

ANSWER: right colic (hepatic) flexure

96. The transverse colon joins the descending colon at the:

ANSWER: left colic (splenic) flexure

97. Within the small intestine, bile salts form water soluble aggregates called:

ANSWER: micelles

98. Lactose is converted into glucose and galactose by the enzyme,:

ANSWER: lactase

99. The lymphatic vessel of an individual villus is called a:
ANSWER: lacteal

TB.26:1

TEST BANK for CHAPTER 26: NUTRITION, METABOLISM, AND THERMOREGULATION

INSTRUCTIONS: ANSWER EACH QUESTION ACCORDING TO THE FOLLOWING KEY:

(A) Only 1 is correct
(B) Only 2 is correct
(C) Both are correct
(D) Neither are correct

1. Carbohydrates are typically ingested as:

    1. polysaccharides
    2. monosaccharides

ANSWER: C

2. Major polysaccharides in foods include:

    1. glycogen
    2. cellulose

ANSWER: C

3. Cellulose:

    1. a fatty acid
    2. makes up plant cell walls

ANSWER: B

4. Anabolic reactions:

    1. simple compounds converted into larger more complex compounds
    2. synthetic aspect of metabolism

ANSWER: C

5. Kilocalorie:

    1. used to measure energy value of food
    2. 1/1000 of a calorie

ANSWER: A

TB.26:2

6. Kilocalorie:

   1. amount of heat required to raise the temperature of one gram of water one degree centigrade
   2. equivalent to 1000 calories

ANSWER: B

7. Common disaccharides consumed:

   1. lactose
   2. sucrose

ANSWER: C

8. The kilocalories yield from 1 gram of carbohydrate is:

   1. 4 kilocalories
   2. equivalent to 20 grams of protein

ANSWER: A

9. Converted to glucose by the liver:

   1. galactose
   2. fructose

ANSWER: C

10. Oxidation:

    1. addition of electrons to a molecule
    2. dehydrogenation

ANSWER: B

11. Oxidation:

    1. removal of hydrogen atoms
    2. removal of electrons

ANSWER: C

12. Oxidation reactions in body:

    1. coupled to reduction reactions
    2. result in loss of energy from substances oxidized

ANSWER: C

13. Redox reactions:

    1. catalyzed by enzymes
    2. are typically uncoupled reactions

ANSWER: A

14. Glycolysis:

    1. does not utilize oxygen
    2. is an anaerobic process

ANSWER: C

15. Aerobic pathway:

    1. electron transport
    2. Kreb's cycle

ANSWER: C

16. Anaerobic pathway:

    1. electron transport
    2. Kreb's cycle

ANSWER: D

17. Anaerobic pathway:

    1. glycolysis
    2. Kreb's cycle

ANSWER: A

18. Aerobic pathway:

    1. electron transport
    2. glycolysis

ANSWER: B

19. Utilizes oxygen:

    1. electron transport
    2. glycolysis

ANSWER: A

TB.26:4

20. Utilizes oxygen:

    1. Kreb's cycle
    2. glycolysis

ANSWER: A

21. Glycolysis:

    1. net gain of two ATP molecules
    2. glucose is phosphorylated

ANSWER: C

22. Glycolysis:

    1. glucose degraded to pyruvic acid
    2. glucose -6-phosphate converted to fructose-6-phosphate

ANSWER: C

23. Tricarboxylic acid cycle:

    1. Kreb's cycle
    2. citric acid cycle

ANSWER: C

24. Electron transport system:

    1. located in the mitochondrial membrane
    2. yields three ATP molecules for each pair of hydrogens entering the system

ANSWER: C

25. Oxygen:

    1. final electron acceptor in electron transport system
    2. is essential for ATP formation via the electron transport system

ANSWER: C

26. Gluconeogenesis:

   1. involves amino acids
   2. formation of 'new' glucose

ANSWER: C

27. Gluconeogenesis is the formation of glucose from:

   1. amino acids
   2. glycerol

ANSWER: C

28. Breakdown of glycogen into glucose:

   1. glycogenesis
   2. glycogenolysis

ANSWER: B

29. Triglycerides may be:

   1. monounsaturated
   2. polyunsaturated

ANSWER: C

30. Triglycerides may contain:

   1. no double bonds
   2. two or more double bonds

ANSWER: C

31. Regarding fatty acids:

   1. most can be synthesized by the body
   2. linoleic acid cannot be synthesized by the body

ANSWER: C

32. Essential fatty acid:

   1. synthesized by the body
   2. linolenic acid

ANSWER: B

33. Linolenic acid:

    1. essential fatty acid
    2. is synthesized by the body

ANSWER: A

34. Essential fatty acid:

    1. linoleic
    2. arachidonic

ANSWER: A

35. Increases one's ch8nces of developing atherosclerosis:

    1. diet high in unsaturated fatty acids
    2. diet low in cholesterol

ANSWER: D

36. Incerases one's chances of developing atherosclerosis:

    1. diet high in saturated fatty acids
    2. diet high in cholesterol

ANSWER: C

37. Chylomicrons:

    1. protein
    2. lipid

ANSWER: C

38. Chylomicrons:

    1. lipoproteins
    2. formed in intestinal epithelial cells

ANSWER: C

39. Low-density lipoprotein (LDL):

    1. produced in liver
    2. contains mainly cholesterol

ANSWER: C

40. Low-density lipoprotein (LDL):

    1. low plasma concentration associated with atherosclerosis
    2. contains hardly any cholesterol

ANSWER: D

41. Low-density-lipoprotein (LDL):

    1. high plasma concentration associated with atherosclerosis
    2. contains mainly cholesterol

ANSWER: C

42. The caloric value of 1 gram of triglyceride:

    1. is equivalent to that of 1 gram of carbohydrate
    2. less than that of 1 gram of carbohydrate

ANSWER: D

43. The caloric value of 1 gram of triglyceride:

    1. is approximately 9 kcal
    2. is more than twice that of a gram of carbohydrate

ANSWER: C

44. The caloric value of 1 gram of triglyceride is more than that for one gram of:

    1. carbohydrate
    2. protein

ANSWER: C

45. Cholesterol is utilized in the formation of:

    1. steroid hormones
    2. bile saltsANSWER: C

46. Ketone body:

    1. acetone
    2. acetoacetic acid

ANSWER: C

47. Ketosis:

    1. abnormally low concentration of ketone bodies in the blood
    2. may lead to death

ANSWER: B

48. Beta oxidation:

    1. breakdown of fatty acids
    2. involves coenzyme A

ANSWER: C

49. Kwashiorkor:

    1. severe protein malnutrition
    2. dehydration, diarrhea, and death may occur

ANSWER: C

50. Phenylketonuria:

    1. an inborn error of metabolism
    2. alanine cannot be converted to phenylalanine

ANSWER: A

51. Phenylketonuria:

    1. phenylalanine cannot be converted to tyrosine
    2. may lead to mental retardation

ANSWER: C

52. Ammonia:

    1. toxic to the body
    2. converted to urea in body

ANSWER: C

53. Negative nitrogen balance:

    1. may occur during starvation
    2. small amounts of protein are degraded

ANSWER: A

54. Positive nitrogen balance:

    1. nitrogen excretion exceeds protein intake
    2. may occur when recovering from an illness

ANSWER: B

55. The amount of nitrogen in the urine:

    1. reflects the amount of amino acid catabolism
    2. equals the amount in the diet when an individual is in nitrogen balance

ANSWER: C

INSTRUCTIONS: COMPLETE THE FOLLOWING:

56. When the intake of protein normally exceeds nitrogen excretion, the individual is said to be in:

ANSWER: positive nitrogen balance

57. When the body excretes more nitrogen than it takes in, it is said to be in:

ANSWER: negative nitrogen balance

58. Trace minerals occur in the body in amounts less than _____ grams.

ANSWER: 5

59. Major minerals are found in the body in amounts greater than _____ grams.

ANSWER: 5

60. Vitamins A, D, E, and K are knowns as the _____ vitamins.

ANSWER: fat soluble

61. Vitamin C belongs to the group of _____ vitamins.

ANSWER: water soluble

62. Triglycerides are broken down in a process called:

ANSWER: lipolysis

63. The appetite center is located in the lateral portion of the:

ANSWER: hypothalamus

64. The satiety center inhibits the _____ center.

ANSWER: appetite

65. The BMR is measured by direct and indirect _____.

ANSWER: calorimetry

66. The amount of energy liberated by the body during metabolism over a finite period of time is called the:

ANSWER: metabolic rate

67. The sum of the BMR and the energy used to carry on all of one's daily activities is called the _____ _____ rate.

ANSWER: total metabolic

68. A person who has an excessive accumulation of body fat and who is more than 20% overweight is said to be:

ANSWER: obese

69. An agent with the potential of producing a fever is called a:

ANSWER: pyrogen

70. A form of starvation most common in children under 1 year of age is known as:

ANSWER: marasmus

INSTRUCTIONS: TRUE / FALSE

71. (T/F): Hormonal problems are the major cause of obesity.

ANSWER: F

72. (T/F): Vasoconstriction of peripheral vessels decreases the loss of heat from the body.

ANSWER: T

73. (T/F): Convection accounts for about 15 % of the body's heat loss.

ANSWER: T

74. (T/F): At room temperature, less than 20% of the body's heat loss occurs by radiation.

ANSWER: F

75. (T/F): Increased heat via increased cellular metabolism is called chemical thermiogenesis.

ANSWER: T

76. (T/F): The liver converts some amino acids into keto acids.

ANSWER: T

77. (T/F): The postabsorptive state is a period of fasting.

ANSWER: T

78. (T/F): The abosorptive state typically lasts for approximately 8 hours.

ANSWER: F

79. (T/F): Complete proteins contain all of the essential
          amino acids.

ANSWER: T

80. (T/F): Incomplete proteins lack one or more of the essential
          amino acids.

ANSWER: T

TB.27:1

TEST BANK for CHAPTER 27: THE URINARY SYSTEM

INSTRUCTIONS: ANSWER EACH QUESTION ACCORDING TO THE FOLLOWING KEY:

    (A) Only 1 is correct
    (B) Only 2 is correct
    (C) Both are correct
    (D) Neither are correct

1. In the clinical condition called renal ptosis the:

    1. kidneys are located at an abnormally low position
    2. ureters may become kinked

ANSWER: C

2. Regarding the renal pyramids:

    1. their apices are called renal papillae
    2. are located only within the cortex

ANSWER: A

3. The renal medulla:

    1. contains all of the glomeruli
    2. where the renal pyramids are located

ANSWER: B

4. The major calyces:

    1. formed by the fusion of several minor calyces
    2. drain into the renal pelvis

ANSWER: C

5. The nephrons:

    1. the functional units of the kidney
    2. more than 6 million per kidney

ANSWER: A

6. The renal corpuscle:

    1. part of a nephron
    2. composed in part of a network of capillaries

ANSWER: C

7. The glomerular capsule:

    1. part of a nephron
    2. part of a renal corpuscle

ANSWER: C

8. The renal corpuscle:

    1. located within the cortical region of the kidney
    2. consists, in part, of Bowman's capsule

ANSWER: C

9. An inflammation of the kidneys involving the glomeruli:

    1. is clinically referred to as glomerulonephritis
    2. is grequently associated with streptococcal infections

ANSWER: C

10. Bright's disease:

    1. is also called glomerulonephritis
    2. may lead to chronic renal disease

ANSWER: C

11. The outer layer of Bowman's capsule:

    1. contains epithelial cells called podocytes
    2. is called the visceral layer

ANSWER: D

12. Podocytes:

    1. are specialized epithelial cells
    2. located on the visceral layer of Bowman's capsule

ANSWER: C

13. Filtration slits:

    1. small spaces between adjacent pedicles
    2. also referred to as slit pores

ANSWER: C

14. Glomerular capillary endothelium:

    1. contains many small fenestrations
    2. forms part of the filtration barrier

ANSWER: C

15. Glomerular basement membrane:

    1. forms part of the filtration barrier
    2. possesses numerous small pores

ANSWER: A

16. The proximal convoluted tubule:

    1. located entirely within the medulla
    2. its wall is composed of pseudostratified columnar epithelium

ANSWER: D

17. The distal convouted tubule:

    1. its wall is composed of simple squamous epithelium
    2. drains contents into collecting tubule

ANSWER: B

18. The juxtaglomerular apparatus:

    1. is responsible for the secretion of renin
    2. consists in part of the macula densa

ANSWER: C

19. Juxtaglomerular cells:

    1. modified smooth muscle cells of the efferent arteriole
    2. specialized cells of the distal convoluted tubule

ANSWER: D

20. The kidneys:

    1. constitute approximately 5% of the total body weight
    2. receive approximately 10% of the cardiac output

ANSWER: D

21. The enzyme, renin:

    1. secreted by the juxtaglomerular apparatus
    2. results in a lowering of renal blood pressure

ANSWER: A

22. Regarding the macula densa:

    1. specialized cells of the afferent arteriole
    2. form part of the juxtaglomerular apparatus

ANSWER: B

23. The vasa recta:

    1. extensions of the efferent arteriole
    2. run perpendicular to the loops of Henle
       and the collecting tubules

ANSWER: A

24. The efferent arteriole:

    1. formed by union of the glomerular capillaries
    2. smaller in diameter than the afferent arteriole

ANSWER: C

25. The efferent arteriole:

    1. receives blood from the glomerular capillaries
    2. connects the glomerular capillaries with the
       peritubular capillaries

ANSWER: C

26. The glomerular filtrate:

   1. in volumetric terms, represents about 20% of the renal blood plasma
   2. is virtually free of proteins

ANSWER: C

27. Influence glomerular filtration:

   1. hydrostatic forces
   2. osmotic forces

ANSWER: C

28. Factors affecting glomerular capillary hydrostatic pressure:

   1. diameter of the efferent arteriole
   2. diameter of the afferent arteriole

ANSWER: C

29. Glomerular capillary hydrostatic pressure:

   1. averages approximately 60 mmHg
   2. is lower than in other systemic capillaries

ANSWER: A

30. Bowman's capsule hydrostatic pressure:

   1. is about 60 mmHg
   2. facilitates filtration

ANSWER: D

31. The plasma oncotic osmotic pressure:

   1. opposes glomerular filtration
   2. averages approximately 75 mmHg

ANSWER: A

32. The net filtration pressure:

   1. influences the glomerular filtration rate (GFR)
   2. is approximately 10 mmHg

ANSWER: C

33. The normal GFR:

    1. is approximately 125 milliliters per minute
    2. reprersents the rate at which water and small solutes flow into Bowman's capsule

ANSWER: C

34. Regarding the GFR:

    1. average urine output typically represents about 1% of the GFR
    2. approximately 99% of the fluid filtered is reabsorbed into the peritubular capillaries

ANSWER: C

35. Increase the effective filtration pressure:

    1. hypoproteinemia
    2. increased permeability of glomerular capillaries

ANSWER: C

36. The enzyme, renin:

    1. converts angiotensinogen to angiotensin I
    2. released by cells of the macula densa

ANSWER: A

37. Regarding angiotensin II:

    1. potent vasodilator
    2. stimulates aldosterone secretion by the adrenal cortex

ANSWER: B

38. Intense sympathetic stimulation:

    1. causes renin to be released by the JG cells
    2. lowers the GFR

ANSWER: C

39. Substances actively reabsorbed by the proximal convoluted tubule:

   1. glucose
   2. bicarbonate

ANSWER: A

40. Passively reabsorbed by the proximal convoluted tubule:

   1. bicarbonate
   2. water

ANSWER: C

INSTRUCTIONS: TRUE - FALSE

41. (T/F): Postassium is actively reabsorbed by the proximal convoluted tubule.

ANSWER: T

42. (T/F): Calcium and phosphate are passively reabsorbed by the proximal convoluted tubule.

ANSWER: T

43. (T/F): Chloride and bicarbonate ions enter the peritubular capillaries by passive diffusion.

ANSWER: T

44. (T/F): Hydrogen and potassium ions are actively secreted into the renal tubule.

ANSWER: F

45. (T/F): Vitamin A deficiency increases one's risk for developing renal calculi or kidney stones.

ANSWER: T

46. (T/F): Hyperparathyroidism may increase one's risk for developing kidney stones or renal calculi.

ANSWER: T

47. (T/F): The vasa recta are important to the integrity of the osmotic gradient of the medulla.

ANSWER: T

48. (T/F): Antidiuretic hormone (ADH) is synthesized by cells of the supraoptic nuclei of the hypothalamus.

ANSWER: T

49. (T/F): Water permeability in the renal collecting tubules is high in the absence of ADH.

ANSWER: F

50. (T/F): Baroreceptors influence ADH-producing cells of the hypothalamus.

ANSWER: T

51. (T/F): Increased blood pressure results in an inhibition of ADH secretion.

ANSWER: T

52. (T/F): The ureter contains a muscular sphincter near its connection with the renal pelvis.

ABSWER: F

53. (T/F): Prostatic hypertrophy may compress the urethra, resulting in dysuria.

ANSWER: T

54. (T/F): Inflammation of the urinary bladder is referred to clinically as cystitis.

ANSWER: T

55. (T/F): The female urethra is embedded within the anterior wall of the vagina.

ANSWR: T

56. (T/F): In the male, the membranous urethra is located between its prostatic and spongy portions.

ANSWER: T

57. (T/F): The membranous portion of the male urethra receives the openings of the ejaculatory ducts.

ANSWER: F

58. (T/F): The spongy urethra in the male is that portion which passes through the urogenital diaphragm.

ANSWER: F

59. (T/F): Stretching of the urinary bladder wall provides a stimulus for initiating the micturition reflex.

ANSWER: T

60. (T/F): The smooth muscle of the urinary bladder is also known as the detrusor muscle.

ANSWER: T

INSTRUCTIONS: COMPLETE THE FOLLOWING:

61. The structure carrying urine from the urinary bladder to the exterior of the body is called the:

ANSWER: urethra

62. Blood may be filtered artificially by a process called:

ANSWER: hemodialysis

63. The clinical term for an inflammation of the renal pelvis and the calyces is:

ANSWER: pyelitis

64. The interlobular arteries of the kidney give rise to the:

ANSWER: afferent arterioles

65. The arcuate arteries gives rise to the:
ANSWER: interlobular arteries

66. The interlobar arteries of the kidney give rise to the:
ANSWER: arcuate arteries

67. The macula densa represents specialized cells of the:
ANSWER: distal convoluted tubule

68. The juxtaglomerular cells are located in the wall of the:
ANSWER: afferent arteriole

69. The enzyme secreted by the juxtaglomerular apparatus is called:
ANSWER: renin

70. Angiotensinogen is converted to angiotensin I by the enzyme,:
ANSWER: renin

71. Solutes leave the ascending vasa recta by diffusion while wat moves in by the process of:
ANSWER: osmosis

72. The rate at which a substance is eliminated from the plasma by the kidneys is called:
ANSWER: plasma clearance

73. Clinically, the presence of erythrocytes in the urine is called:
ANSWER: hematuria

74. The presence of glucose in the urine is called:
ANSWER: glycosuria

75. Fluid gains that decrease extracellular osmolarity and decrease ADH secretion lead to a form of increased urine output called:

ANSWER: water diuresis

76. Renal tubules containing excessive amounts of unreabsorbed solutes may result in an increased urine output designated as:

ANSWER: osmotic diuresis

77. A sunstance used clinically to measure plasma clearance:

ANSWER: inulin

78. Vessels which are tributaries of the arcuate veins are the:

ANSWER: interlobar veins

79. The kidneys are located posterior to the peritoneum and are therefore said to be:

ANSWER: retroperitoneal

80. The indentation along the medial border of the kidney is called the:

ANSWER: hilus

TB.28:1

TEST BANK for CHAPTER 28:FLUID, ELECTROLYTE, AND
                       ACID-BASE BALANCE

INSTRUCTIONS: ANSWER EACH QUESTION ACCORDING TO THE FOLLOWING KEY:

           (A) Only 1 is correct
           (B) Only 2 is correct
           (C) Both are correct
           (D) Neither are correct

1.   Water:

        1. universal biological solvent
        2. nonpolar molecule

ANSWER: A

2.   Water:

        1. polar molecule
        2. universal biological solvent

ANSWER: C

3.   Water:

        1. acts to minimize temperature changes
           throughout the body
        2. accounts for approximately 20% of the adult
           human body weight

ANSWER: A

4.   Water as percent of body weight:

        1. lower in infants than in adults
        2. lower in women than in men

ANSWER: B

5.   Water content:

        1. less in fat than other tissues
        2. varies in different tissues

ANSWER: C

6. Source of water loss:

    1. feces
    2. lungs

ANSWER: C

7. Source of water loss:

    1. skin
    2. urine

ANSWER: C

8. Major fluid compartment of body:

    1. interstitial fluid
    2. intracellular fluid

ANSWER: C

9. Major fluid compartment of body:

    1. blood plasma
    2. interstitial fluid

ANSWER: C

10. Component of extracellular fluid:

    1. interstitial fluid
    2. blood plasma

ANSWER: C

11. Component of extracellular fluid:

    1. cerebrospinal fluid
    2. synovial fluid

ANSWER: C

12. Blood plasma accounts for approximately:

    1. 5% of the body weight
    2. 50% of the extracellular fluid

ANSWER: A

TB.28:3

13. Blood plasma accounts for approximately:

   1. 20% of the body weight
   2. 5% of the extracellular fluid

ANSWER: D

14. The interstitial fluid accounts for approximately:

   1. 30% of the body weight
   2. 10% of the extracellular fluid

ANSWER: D

15. The interstitial fluid accounts for approximately:

   1. 15% of the body weight
   2. 80% of the extracellular fluid

16. Force favoring movement of water from blood
    to interstitial fluid:

   1. capillary hydrostatic pressure
   2. capillary osmotic pressure

ANSWER: A

17. Force favoring movement of water from blood
    to interstitial fluid:

   1. capillary osmotic pressure
   2. interstitial fluid osmotic pressure

ANSWER: B

18. Force favoring movement of water from blood
    to interstitial fluid:

   1. interstitial fluid osmotic pressure
   2. capillary hydrostatic pressure

ANSWER: C

19. Intracellular fluid accounts for approximately:

   1. 40% of the body weight
   2. 63% of the total body water content

ANSWER: C

20. Intracellular fluid accounts for approximately:

    1. 10% of the body weight
    2. 63 % of the total body water

ANSWER: B

21. Intracellular fluid accounts for approximately:

    1. 40% of the body weight
    2. 15% of the total body water

ANSWER: A

22. Antidiuretic hormone (ADH):

    1. regulates body fluid volume
    2. regulates extracellular osmolarity

ANSWER: C

23. Thirst:

    1. major regulator of water intake
    2. has regulatory center in medulla

ANSWER: A

24. Thirst:

    1. major regulator of water intake
    2. has regulatory center in hypothalamus

ANSWER: C

25. Edema:

    1. reduced amount of interstitial fluid
    2. may result from high blood pressure

ANSWER: A

26. Edema:

    1. excess interstitial fluid
    2. uncommon in lymphatic blockage

ANSWER: A

27. Edema:

    1. common in lymphatic blockage
    2. excess interstitial fluid

ANSWER: C

28. Edema:

    1. body fluid depletion
    2. may result from high blood pressure

ANSWER: B

29. Dehydration:

    1. body fluid depletion
    2. may be hypertonic or isotonic in nature

ANSWER: C

30. Cause of isotonic dehydration:

    1. excessive vomiting
    2. burns

ANSWER: C

31. Cause of isotonic dehydration:

    1. hemorrhage
    2. diarrhea

ANSWER: C

32. Cause of hypertonic dehydration:

    1. excessive sweating
    2. burns

ANSWER: C

33. Overhydration:

    1. water intoxication
    2. may result from excessive sweating

ANSWER: C

34. Electrolyte:

   1. dissociates into ions in solution
   2. cannot conduct and electrical current

ANSWER: A

35. Electrolyte:

   1. conducts an electrical current
   2. dissociates into ions in solution

ANSWER: C

36. Sodium ion:

   1. major anion of the blood plasma
   2. freely filtered at glomerulus

ANSWER: B

37. Sodium ion:

   1. major cation of blood plasma
   2. higher extracellular concentration than intracellular concentration

ANSWER: C

38. Aldosterone:

   1. produced by adrenal medulla
   2. decreases sodium reabsorption from distal convoluted tubule

ANSWER: D

39. Aldosterone:

   1. produced by adrenal cortex
   2. functions in concert with ADH to regulate fluid output from the body

ANSWER: C

40. Atrial natriuretic factor:

    1. inhibits renin release
    2. inhibits aldosterone release

ANSWER: C

41. Atrial natriuretic factor stimulates:

    1. renin release
    2. aldosterone release

ANSWER: D

42. Potassium:

    1. major cation of intracellular fluid
    2. intracellular concentration is higher than extracellular concentration

ANSWER: C

43. Phosphate ion:

    1. most abundant extracellular anion
    2. required for nucleic acid synthesis

ANSWER: B

44. Phosphate ion:

    1. most abundant intracellular anion
    2. required for synthesis of high-energy compounds

ANSWER: C

45. Parathyroid hormone:

    1. released when blood calcium levels fall
    2. stimulates release of phosphate from bones

ANSWER: C

46. Parathyroid hormone:

    1. released when blood calcium levels increase
    2. stimulates release of calcium from bones

ANSWER: B

47. Calcitonin:

    1. released in response to increased blood calcium level
    2. inhibits release of phosphate from bone

ANSWER: C

48. Calcitonin:

    1. inhibits absorption of calcium from gastrointestinal tract
    2. increases amount of calcium excreted by kidney

ANSWER: C

49. The normal pH of arterial blood:

    1. is 7.4
    2. is lower than that of venous blood

ANSWER: A

50. Occurs when hydrogen ion concentration of arterial blood increases:

    1. alkalosis
    2. pH increases

ANSWER: D

51. Occurs when hydrogen ion concentration of arterial blood increases:

    1. acidosis
    2. pH decreases

ANSWER: C

52. Occurs when hydrogen ion concentration of arterial blood decreases:

    1. acidosis
    2. pH decreases

ANSWER: D

53. Occurs when hydrogen ion concentration of arterial blood decreases:

    1. alkalosis
    2. pH increases

ANSWER: C

INSTRUCTIONS: COMPLETE THE FOLLOWING:

54. The primary cause of respiratory acidosis is:

ANSWER: hypoventilation

55. The primary cause of respiratory alkalosis is:

ANSWER: hyperventilation

56. Clinically, the presence of excess sodium ions is called:

ANSWER: hypernatremia

57. Clinically, the presence of excess potassium ions is called:

ANSWER: hyperkalemia

58. Clinically, a deficiency in potassium ions is called:

ANSWER: hypokalemia

59. A chemical that dissociates into ions in solution and can conduct an electrical current is called an:

ANSWER: electrolyte

60. Positively charged ions are called:

ANSWER: cations

61. Negatively charged ions are called:

ANSWER: anions

62. Extreme overhydration is also known as:

ANSWER: water intoxication

63. When water is lost from the body at a higher rate than are electrolytes, the condition is called:

ANSWER: hypertonic dehydration

64. When there is no noticable decrease in electrolyte concentration with fluid loss, the condition is called:

ANSWER: isotonic dehydration

65. When both fluid and electrolyte levels become depressed, the condition is called:

ANSWER: hypotonic dehydration

66. A condition in which there is an excess of fluid in the interstitial compartment is known as:

ANSWER: edema

67. Water and dissolved substances constitute the:

ANSWER: body fluid

68. The fluid compartment that constitutes about 40% of the body weight is the:

ANSWER: intracellular fluid

69. The fluid compartment that accounts for nearly 63% of the total body water is the:

ANSWER: intracellular fluid

70. Water is known as the universal biological:

ANSWER: solvent

71. Water added to the body as a result of various catabolic reactions is called:

ANSWER: metabolic water

72. Aldosterone secretion is controlled by a complex mechanism called the:

ANSWER: renin-angiotensin-aldosterone pathway

73. Thyroid hormone inhibiting the release of calcium and phosphate from bone is:

ANSWER: calcitonin

74. A solution of two or more chemicals that prevent a marked change in the hydrogen ion concentration when either an acid or base is added is called a:

ANSWER: buffer

75. A condition of increased pH is called:

ANSWER: acidosis

TB.29:1

TEST BANK for CHAPTER 29: REPRODUCTION

INSTRUCTIONS: ANSWER EACH QUESTION ACCORDING TO THE FOLLOWING KEY:

    (A) Only 1 is correct
    (B) Only 2 is correct
    (C) Both are correct
    (D) Neither are correct

1. Metaphase I:

    1. chromosomes line up on equator of cell
    2. homologous chromosomes, separate

ANSWER: A

2. Anaphase I:

    1. synapsis occurs
    2. mitotic spindle forms

ANSWER: D

3. Meiosis II:

    1. four haploid cells from each primary sex cell
    2. permits chromatids to separate

ANSWER: C

4. Testis:

    1. covered with capsule called tunica albuginea
    2. contains seminiferous tubules

ANSWER: C

5. Spermatogenesis:

    1. spermiogenesis is its final stage
    2. typically requires more than 100 days

ANSWER: A

TB.29:2

6. Spermatogonia:

   1. develop from spermatids
   2. multiply by meiosis

ANSWER: D

7. Primary spermatocytes:

   1. multiply by meiosis
   2. differentiate from spermatogonia

ANSWER: B

8. Secondary spermatocytes:

   1. arise through meiosis from primary spermatocytes
   2. have haploid number of chromosomes

ANSWER: C

9. Spermatids:

   1. formed during meiosis II
   2. four arise from each secondary spermatocyte

ANSWER: A

10. Spermatids:

    1. each has diploid number of chromosomes
    2. become spermatozoa during spermiogenesis

ANSWER: B

11. Sertoli cells:

    1. provide nourishment to the developing gametes
    2. secrete inhibin

ANSWER: C

12. Inhibin:

    1. secreted by secondary sprematocytes
    2. regulates secretion of LH

ANSWER: D

13. Inhibin:

    1. secreted by anterior lobe of pituitary gland
    2. regulates FSH secretion

ANSWER: B

14. Inhibin:

    1. secretd by Sertoli cells
    2. regulates secretion of FSH

ANSWER: C

15. Sertoli cells:

    1. their bases are joined by tight junctions
    2. form blood-testis barrier

ANSWER: C

16. Acrosome:

    1. modified Golgi body
    2. aids fertilzation with its enzymes

ANSWER: C

17. Components of a mature sperm cell:

    1. midpiece
    2. raphe

ANSWER: A

18. Features of a mature sperm cell:

    1. acrosome
    2. midpiece

ANSWER: C

19. Inguinal hernia:

    1. failure of testis to descend through inguinal canal
    2. also called cryptorchism

ANSWER: D

20. Cryptorchism:

    1. may lead to sterility if left untreated
    2. may be treated with gonadotropic hormones

ANSWER: C

21. Ejaculatory duct:

    1. empties its contents into the seminal vesicle
    2. formed by vas deferens and duct of seminal vesicle

ANSWER: B

22. Ejaculatory duct:

    1. opens into prostatic portion of urethra
    2. is not a paired structure

ANSWER: A

23. Semen:

    1. secretions from bulbourethral glands
    2. secretions from prostate gland

ANSWER: C

24. Seminal vesicles:

    1. secretions account for less than 10% of semen volume
    2. secretions contain the sugar fructose

ANSWER: B

25. Prostate gland:

    1. is a paired gland
    2. receives contents of ejaculatory ducts

ANSWER: B

26. Prostate gland:

    1. located on inferior aspect of urinary bladder
    2. its tissue surrounds the urethra

ANSWER: C

27. Sperm:

   1. account for less than 1% of ejaculate volume
   2. numbers increase with repeated ejaculation

ANSWER: A

28. Characteristic feature of the penis:

   1. corona
   2. frenulum

ANSWER: C

29. Corpus spongiosum:

   1. a paired structure
   2. distal end forms glans penis

ANSWER: B

30. Erectile tissue:

   1. corpus cavernosa
   2. corpus spongiosum

ANSWER: C

31. Erection may be initiated by:

   1. tactile stimulation
   2. erotic thoughts

ANSWER: C

32. Chronic inability to sustain an erection:

   1. erectile dysfunction
   2. impotence

ANSWER: C

33. Reflex action:

   1. ejaculation
   2. erection

ANSWER: C

34. Leydig cells:

   1. secrete inhibin
   2. interstitial cells

ANSWER: C

35. Leydig cells:

   1. secrete androgens
   2. located in seminal vesicles

ANSWER: A

36. Produce estrogens:

   1. prostate gland
   2. seminal vesicles

ANSWER: C

37. The hormone inhibin:

   1. produced by cells of seminiferous tubules
   2. inhibits synthesis of FSH by anterior pituitary gland

ANSWER: A

38. Testosterone:

   1. responsible for adult male primary sexual characteristics
   2. responsible for male secondary sexual characteristics

ANSWER: C

39. The ovary:

   1. is a paired structure
   2. secretes estrogen

ANSWER: C

40. The ovaries:

   1. secrete progesterone
   2. held in position by connective tissue ligaments

ANSWER: C

41. The ovary:

    1. covered by tunica albuginea
    2. has cortex and medulla

ANSWER: C

42. Follicle:

    1. ovum plus its granulosa cells
    2. some become atretic

ANSWER: C

43. Granulosa cells:

    1. secrete fluid
    2. form a portion of a follicle

ANSWER: C

44. Graafian follicle:

    1. located deep within ovarian medulla
    2. an immature follicle

ANSWER: D

45. Zona pellucida:

    1. a thick membrane
    2. formed by secretions of granulosa cells

ANSWER: C

46. Salpingitis:

    1. inflammation of the ovary
    2. typically caused by gonorrheal infection

ANSWER: B

47. Feature of uterine tube:

    1. fimbriae
    2. corpus albicans

ANSWER: A

48. Feature of the uterine tube:

    1. infundibulum
    2. fimbriae

ANSWER: C

49. Uterine tubes:

    1. lined by ciliated mucous membrane
    2. has free end, the infundibulum

ANSWER: C

50. Uterine tube:

    1. typically the site of fertilization
    2. has an intramural portion

ANSWER: C

INSTRUCTIONS: COMPLETE THE FOLLOWING:

51. A pregnancy in which the embryo begins to develop outside of the uterus is called an:

ANSWER: ectopic pregnancy

52. Inflammation of the uterine tube is called:

ANSWER: salpingitis

53. Ejection of the secondary oocyte from the ovary is called:

ANSWER: ovulation

54. The middle layer of the uterine wall is called the:

ANSWER: myometrium

55. The innermost lining of the uterus is designated as the:

ANSWER: endometrium

56. The superficial layer of the endometrium is called the:

ANSWER: functional layer

57. The rounded portion of the uterus above the level of the entrance of the uterine tubes is known as the:

ANSWER: fundus

58. The double fold of peritoneum which attaches the uterus to the pelvis walls is called the:

ANSWER: broad ligament

59. Recesses formed between the vaginal wall and the cervix of the uterus are called:

ANSWER: fornices

60. About 95% of cervical cancer cases are of what type?

ANSWER: squamous cell carcinomas

61. An abnormal posterior tipping of the uterus is known clinically as:

ANSWER: retroflexion

62. The lower narrow portion of the uterus is called the:

ANSWER: cervix

63. The junction of the uterine cavity with the cervical canal is designated as the:

ANSWER: internal os

64. The small body of erectile tissue homologous to the male glans penis is the:

ANSWER: clitoris

65. Bartholin's glands are also known as:

ANSWER: greater vestibular glands

66. The exposed portion of the clitoris is called the:

ANSWER: glans

67. The labia minora merge anteriorly to form the _____ of the clitoris.

ANSWER: prepuce

68. The anatomical space enclosed by the labia minor is the:

ANSWER: vestibule

69. The diamond-shaped region located between the anus and the pubic arch is called the:

ANSWER: perineum

70. Ligaments of Cooper firmly attach the _____ to the skin.

ANSWER: breasts

INSTRUCTIONS:TRUE / FALSE

71. (T/F): The mammary gland is synonymous with breast.

ANSWER: F

72. (T/F): The mammary glands are located within the breast.

ANSWER: T

73. (T/F): The clinical (gynecological) perineum lies between the anus and the vagina.

ANSWER: T

74. (T/F): Mastectomy is the term applied to surgical removal of the breast.

ANSWER: T

75. (T/F): Progesterone stimulates uterine contractions.

ANSWER: F

76. (T/F): Testosterone is found in males and females.

ANSWER: T

77. (T/F): Ovulation typically occurs about 14 days prior to the beginning of the next menstrual cycle.

ANSWER: T

78. (T/F): The postovulatory phase of the menstrual cycle is also called the luteal, or secretory phase.

ANSWER: T

79. (T/F): Function of the corpus luteum is sustained by the secretion of HCG from the ovary.

ANSWER: F

80. (T/F): Absence of menstruation is referred to clinically as amenorrhea.

ANSWER: T

81. (T/F): Following ejaculation, sperm remain viable only for 24 hours.

ANSWER: F

82. (T/F): Vasectomy does not affect masculinity.

ANSWER: T

83. (T/F): Vasectomy typically results in a marked reduction in the volume of semen ejaculated at orgasm.

ANSWER: F

84. (T/F): Gonorrhea, if untreated, may affect the heart valves and meninges.

ANSWER: T

85. (T/F): Only females may contract gonorrhea.

ANSWER: F

86. (T/F): Gummas are characteristic of advanced syphilis.

ANSWER: T

87. (T/F): Genital herpes is caused by herpes simplex virus type 2.

ANSWER: T

88. (T/F): If treated early, genital herpes can be completely cured with penicillin.

ANSWER: F

TEST BANK for CHAPTER 30: DEVELOPMENT

INSTRUCTIONS: COMPLETE THE FOLLOWING:

1. The zygote is covered by a membranous layer called the:
ANSWER: zona pellucida

2. As the zygote undergoes mitosis, a solid mass of cells termed the _____ results.
ANSWER: morula

3. The mitotic divisions of the zygote are collectively termed:
ANSWER: cleavage

4. The first cleavage division typically requires a time interval of approximately _____ for completion.
ANSWER: 24 hours

5. The zygote is ordinarily formed in the:
ANSWER: uterine tube

6. The morula is typically converted into a structure called the:
ANSWER: blastocyst

7. The peripheral layer of the blastocyst is referred to as the:
ANSWER: trophoblast

8. The trophoblastic cells participate in the formation of the chorion and the:
ANSWER: placenta

9. The blastocyst contains an inner group of cells called the:
ANSWER: inner cell mass

10. Ectoderm, endoderm, and mesoderm are known as the three primary:

ANSWER: germ layers

11. By the end of the fourth week of human development, the neural plate has been converted into the:

ANSWER: neural tube

12. The end of the second month of human intrauterine development marks the end of the _____ period.

ANSWER: embryonic

13. The process by which the baby is expelled from the uterus is called:

ANSWER: parturition

14. Stretching of the uterine cervix results in the release of _____ from the posterior pituitry gland.

ANSWER: oxytocin

15. Surgical incision of the perineal region to facilitate the expulsion of the baby is called an:

ANSWER: episiotomy

16. Once delivered, the placenta together with the umbilical cord and fetal membranes is called the:

ANSWER: afterbirth

17. The third stage of labor is also referred to as the:

ANSWER: placental stage

18. Together the trophoblast and the extraembryonic somatic mesoderm represent the:

ANSWER: chorion

19. Cells that line the inner aspect of the amniotic cavity are termed the:

ANSWER: amnion

20. Fertilization typically occurs within the:

ANSWER: uterine tube

INSTRUCTIONS:TRUE / FALSE

21. (T/F): The blastocyst contains a fluid filled cavity.

ANSWER: T

22. (T/F): The human ovum contains yolk.

ANSWER: F

23. (T/F): Cells of the syncytiotrophoblast release lytic enzymes.

ANSWER: T

24. (T/F): The cytotrophoblast eventually surrounds the entire trophoblast.

ANSWER: T

25. (T/F): Only the secondary yolk sac in humans contains yolk.

ANSWER: F

26. (T/F): Nutrition and excretion are functions attributed to to the placenta.

ANSWER: T

27. (T/F): Protection and hormone secretion are functions attributed to the placenta.

ANSWER: T

28. (T/F): Chorionic villi become vascularized as the embryonic circulatory system becomes established.

ANSWER: T

29. (T/F): The allantois is a vestigial structure in humans.

ANSWER: T

30. (T/F): The end of the eighth week of intrauterine development marks the beginning of the fetal period.

ANSWER: T

31. (T/F): The developing human is most susceptible to teratogenic insult during the second trimester of pregnancy.

ANSWER: F

32. (T/F): The developing human is equally susceptible to teratogenic insult in all three trimesters of pregnancy.

ANSWER: F

33. (T/F): Phocomelia is a typical malformation observed in thalidomide teratogenesis.

ANSWER: T

34. (T/F): The baby is delivered in the second stage of labor.

ANSWER: T

35. (T/F): The first stage of labor is typically the shortest.

ANSWER: F

36. (T/F): The uterine cervix becomes dilated and effaced during the first stage of labor.

ANSWER: T

37. (T/F): Oxytocin inhibits contractions of uterine smooth muscle.

ANSWER: F

38. (T/F): Estrogen markedly stimulates uterine contractility.

ANSWER: T

39. (T/F): Progesterone stimulates uterine contractility.

ANSWER: F

40. (T/F): Rupture of the amnion may occur during the first stage of labor.

ANSWER: T

41. (T/F): "Breaking of the bag of waters" refers to rupture of the amnion.

ANSWER: T

42. (T/F): Episiotomy facilitates expulsion of the baby and the tearing of maternal tissue.

ANSWER: T

43. (T/F): Lanugo is evident during the first month of embryonic development.

ANSWER: F

44. (T/F): The notochord plays a role in the induction of cells which form the neural plate.

ANSWER: T

45. (T/F): In the developing human, the fetal period and embryonic period are of equivalent length.

ANSWER: F

46. (T/F): Chorionic villus sampling is a procedure utilized for prenatal diagnosis.

ANSWER: T

47. (T/F): A clinical report for amniocentesis requires two or three weeks for completeion whereas that for CVS may be obtained within hours.

ANSWER: T

48. (T/F): Cells of the trophoblast secrete HCG.

ANSWER: T

49. (T/F): Cleavage is the collectively term applied to the mitotic divisions of the zygote.

ANSWER: T

50. (T/F): The morula typically is composed of 20 to 30 cells.

ANSWER: F

51. (T/F): The zona pellucida is a glycoprotein-rich membranous layer surrounding the zygote.

ANSWER: T

52. (T/F): The morula is markedly larger than the zygote.

ANSWER: F

53. (T/F): Cells of the trophoblast participate in the formation of the placenta.

ANSWER: T

54. (T/F): Cells of the trophoblast do not participate in the formation of the chorion.

ANSWER: F

55. (T/F): Once formed, the blastocyst floats free within the uterine cavity for about six days.

ANSWER: F

56. (T/F): Lacunae develop within the syncytiotrophoblast and become filled with maternal blood.

ANSWER: T

57. (T/F): A tubal pregnancy represents an ectopic pregnancy.

ANSWER: T

58. (T/F): The developing fetus may swallow amniotic fluid from time to time later in pregnancy.

ANSWER: T

59. (T/F): Mesoderm covering the primary yolk sac is referred to as the extraembryonic splanchnic mesoderm.

ANSWER: T

60. (T/F): The yolk sac is thought to be a vestigial structure in humans.

ANSWER: T

61. (T/F): Pharyngeal arches are separated from one another internally by pharyngeal grooves.

ANSWER: F

62. (T/F): Lung buds arise from the pharyngeal region of the embryonic foregut.

ANSWER: T

63. (T/F): The rudiments of all future organs have been developed by the end of the second month of gestation in humans.

ANSWER: T

64. (T/F): Blastocyst implantation typically starts on the second day following fertilization.

ANSWER: F

65. (T/F): The foramen ovale represents a fetal shunt.

ANSWER: T

66. (T/F): The ductus arteriosus is a fetal shunt that later becomes the ligamentum arteriosum.

ANSWER: T

67. (T/F): Birth defects may be due to genetic factors, environmental factors, or a combination of the two.

ANSWER: T

68. (T/F): Ingestion of alcohol during pregnancy is known to be developmentally toxic.

ANSWER: T

69. (T/F): Heroin use during pregnancy is associated with a high prematurity rate.

ANSWER: T

70. (T/F): Rubella virus infection during the first month of pregnancy may result in deafness and heart defects of the newborn.

ANSWER: T

71. (T/F): Pregnant women are routinely tested for syphilis during prenatal examinations.

ANSWER: T

72. (T/F): Ultrasonography may be utilized for prenatal diagnosis.

ANSWER: T

73. (T/F): The neonatal time period extends from birth until the third postnatal month.

ANSWER: F

74. (T/F): Progeria is a genetic abnormality resulting in premature aging of the individual.

ANSWER: T

75. (T/F): Individuals with progeria typically die of heart disease related to atherosclerosis by the age of 10 to 15 years.

ANSWER: T

TB.31:1

TEST BANK for CHAPTER 31:INHERITANCE

INSTRUCTIONS: COMPLETE THE FOLLOWING:

1. Each triplet or codon specifies a particular:

ANSWER: amino acid

2. The sugar in the backbone of DNA is:

ANSWER: deoxyribose

3. Together a sugar, phosphate group, and a base constitute a unit called a:

ANSWER: nucleotide

4. The sex cells are also called the:

ANSWER: gametes

5. Gametes are formed by a special type of cell division called:

ANSWER: meiosis

6. The branch of biology concerned with heredity is called:

ANSWER: genetics

7. Each gamete contains the _____ number of chromosomes.

ANSWER: haploid

8. The genetically active strand of DNA is called the _____ strand.

ANSWER: coding or plus

9. Process by which DNA segment is copied in the form of a messenger RNA molecule is called:

ANSWER: transcription

10. The sugar in the RNA backbone is called:

ANSWER: ribose

11. The conversion of information in mRNA into a polypeptide is called:

ANSWER: translation

12. Anticodons are associated with which type of RNA?

ANSWER: transfer RNA (tRNA)

13. Agents responsible for genetic mutations are called:

ANSWER: mutagens

14. Two genes governing the same trait are designated as:

ANSWER: alleles

15. Genes that are located near one another on the same chromosome and do not undergo independent assortment are said to be:

ANSWER: linked

16. Individuals unable to synthesize the pigment, melanin, are said to have a condition called:

ANSWER: albinism

17. A cross between two individuals that differ with regard to the alleles they carry for a single locus is designated as a:

ANSWER: monohybrid cross

18. The genetic makeup of an individual is referred to as its:

ANSWER: geneotype

19. The term _____ refers to how the genes are expressed.

ANSWER: phenotype

20. In albinism, as in most genetic disorders, the normal gene is the:

ANSWER: dominant gene

INSTRUCTIONS: TRUE/FALSE

21. (T/F): Typically, the recessive gene is expressed only in the homozygous condition.

ANSWER: T

22. (T/F): Genes may be codominant.

ANSWER: T

23. (T/F): Phenylketonuria (PKU) is treated clinically by a special diet that is low in phenylalanine.

ANSWER: T

24. (T/F): The karyotype for Klinefelter syndrome is XXX.

ANSWER: F

25. (T/F): The karyotype for Down syndrome is trisomy 21.

ANSWER: T

26. (T/F): Cri-du-chat syndrome involves deletion of the short arm of chromosome 5.

ANSWER: T

27. (T/F): Cystic fibrosis is an autosomal recessive disorder.

ANSWER: T

28. (T/F): Tay-Sachs disease is an X-linked recessive disorder.

ANSWER: F

29. (T/F): Lesch-Nyhan syndrome is an X-linked recessive disorder.

ANSWER: T

30. (T/F): A codon of DNA specifies a particular amino acid.
ANSWER: T

31. (T/F): Each human somatic cell contains 46 pairs of chromosomes.
ANSWER: F

32. (T/F): Each human somatic cell contains 23 pairs of chromosomes.
ANSWER: T

33. (T/F): Human somatic cells contain the haploid number of chromosomes.
ANSWER: F

34. (T/F): Human somatic cells contain the diploid number of chromosomes.
ANSWER: T

35. (T/F): Each human gamete contains 23 pairs of chromosomes.
ANSWER: F

36. (T/F): Each human agmete contains 23 chromosomes.
ANSWER: T

37. (T/F): The zygote contains the diploid number of chromosomes.
ANSWER: T

38. (T/F): The zygote contains the haploid number of chromosomes.
ANSWER: F

39. (T/F): Deoxyribose is a pentose sugar.
ANSWER: T

40. (T/F): The two strands of DNA are attached to one another by hydrogen bonds.

ANSWER: T

41. (T/F): If you know the base sequence of one of the DNA strands, you can predict te base sequence of the other.

ANSWER: T

42. (T/F): Adenine of one DNA strand will only pair with uracil of the other strand.

ANSWER: F

43. (T/F): Each strand of DNA replicates prior to mitosis.

ANSWER: T

44. (T/F): Replication of DNA proceeds from the 5′ to the 3′ end of the molecule.

ANSWER: F

45. (T/F): Replication of DNA proceeds from the 3′ to the 5′ end of the molecule.

ANSWER: T

46. (T/F): Messenger RNA contains the information specifying the amino acid sequence of a polypeptide chain.

ANSWER: T

47. (T/F): All forms of RNA are double stranded molecules like DNA.

ANSWER: F

48. (T/F): Thymine substitutes for uracil as a base in RNA.

ANSWER: F

49. (T/F): Uracil substitutes for thymine as a base in RNA.

ANSWER: T

50. (T/F): Genetic information is coded in mRNA in sets of three bases, the codons.

ANSWER: F

51. (T/F): Each kind of tRNA associates with and transfers only one kind of amino acid.

ANSWER: T

52. (T/F): Anticodons on tRNA recognize codons on mRNA by complementary base pairing.

ANSWER: T

53. (T/F): A 'nonsense' sequence is a codon that has no amino acid equivalent.

ANSWER: T

54. (T/F): A abrupt change in the genetic information of a cell is called a mutation.

ANSWER: T

55. (T/F): Mutations are harmful in that they provide genetic variation within a species.

ANSWER: F

56. (T/F): Many mutagens are also considered to be carcinogenic.

ANSWER: T

57. (T/F): The principle of independent assortment states that alleles of different loci assort randomly into the gametes during meiosis.

ANSWER: T

58. (T/F): Inheritance of gender is a good example of separation of chromosomes during meiosis.

ANSWER: T

TB.31:7

59. (T/F): The 22 pairs of matched chromosomes are known as the autosomes.

ANSWER: T

60. (T/F): The pair of sex chromosomes are alike in males and different in females.

ANSWER: F

INSTRUCTIONS: ANSWER EACH QUESTION ACCORDING TO THE FOLLOWING KEY:

(A) Only 1 is correct
(B) Only 2 is correct
(C) Both are correct
(D) Neither are correct

61. A mutation:

1. represents an abrupt change in genetic information of the cell
2. does not occur spontaneously

ANSWER: A

62. Ribosomal RNA:

1. is necessary for protein synthesis
2. is required for transcription

ANSWER: A

63. Anticodons:

1. are located on ribosomal RNA
2. recognize codons on mRNA

ANSWER: B

64. Transfer RNA:

1. is required for translation to occur
2. possesses anticodons

ANSWER: C

65. Observed in alkaptonuria:

    1. pigmentation of cartilage
    2. urine darkens upon standing

ANSWER: C

66. Gangliosidosis:

    1. several types exist
    2. cystic fibrosis is one type

ANSWER: A

67. Tay-Sachs disease:

    1. is a gangliosidosis
    2. especially prevalent among Jews of Eastern European ancestry

ANSWER: C

68. Multiple congenital defects are seen in:

    1. Trisomy 13
    2. Trisomy 15

ANSWER: C

69. Down syndrome:

    1. Trisomy 18
    2. mongolism

ANSWER: B

70. Multiple alleles:

    1. ABO blood types
    2. ocuur with a uniform frequency within a population

ANSWER: A

TB.32:1

TEST BANK for CHAPTER 32: FUNDAMENTALS OF EXERCISE PHYSIOLOGY

INSTRUCTIONS: ANSWER EACH QUESTION ACCORDING TO THE FOLLOWING KEY:

   (A) Only 1 is correct
   (B) Only 2 is correct
   (C) Both are correct
   (D) Neither are correct

1. Maximal oxygen consumption:

   1. varies with lifestyle
   2. declines with advancing age

ANSWER: C

2. Maximal oxygen consumption:

   1. may be increased with training
   2. is not influenced by lifestyle

ANSWER: A

3. Maximal oxygen consumption:

   1. a critical factor in one's capacity for extended vigorous physical activity
   2. increases as one further increases the level of physical activity

ANSWER: A

4. Maximal oxygen consumption is influenced by:

   1. gender
   2. age

ANSWER: C

5. Maximal oxygen consumption:

   1. rises markedly until we enter our late teens
   2. declines steadily after the age of 25

ANSWER: C

6. Oxygen debt following moderate exercise:

    1. fast component
    2. lactic acid oxygen debt

ANSWER: A

7. Oxygen debt following moderate exercise:

    1. slow component
    2. fast component

ANSWER: B

8. Oxygen debt following strenuous exercise:

    1. alactic acid oxygen debt
    2. slow component

ANSWER: B

9. Oxygen debt following strenuous exercise:

    1. fast component lasts for several minutes
    2. slow component is called lactic acid oxygen debt

ANSWER: C

10. Aerobic exercise continued for long intervals:

    1. lactic acid accumulates
    2. recovery requires longer time than moderate exercise

ANSWER: C

11. Minute respiratory volume increases with:

    1. rate of breathing
    2. depth of breathing

ANSWER: C

TB.32:3

12. Rises during exercise:

    1. rate of alveolar diffusion for oxygen
    2. rate of alveolar diffusion for carbon dioxide

ANSWER: C

13. Changes with exercise:

    1. distribution of blood flow
    2. rate of blood flow to muscles

ANSWER: C

14. Exhibits decreased rate of blood flow in strenuous exercise:

    1. brain
    2. gastrointestinal tract

ANSWER: B

15. Exhibits decreased rate of blood flow in strenuous exercise:

    1. kidneys
    2. gastrointestinal tract

ANSWER: C

16. Exhibits increased rate of blood flow in strenuous exercise:

    1. coronary arteries
    2. skeltal muscle

ANSWER: C

17. Exhibits increased rate of blood flow in strenuous exercise:

    1. kidneys
    2. coronary arteries

ANSWER: B

18. Prolonged athletic training often leads to:

    1. cardiac hypertrophy
    2. increased stroke volume

ANSWER: C

19. Cardiac output is influenced by:

    1. venous return to the heart
    2. degree of physical activity

ANSWER: C

20. Frequently experienced by females undertaking excessive exercise training:

    1. oligomenorrhea
    2. amenorrhea

ANSWER: C

INSTRUCTIONS: TRUE / FALSE

21. (T/F): Lymphocytosis may result from strenuous physical activity of short duration.

ANSWER: T

22. (T/F): Cardiac output is influenced by the venous return to the heart.

ANSWER: T

23. (T/F): Antidiuretic hormone is released in increased amounts during physical exercise.

ANSWER: T

24. (T/F): Release of growth hormone is not influenced by physical activity.

ANSWER: F

25. (T/F): Increasing levels of physical activity lead to decreased secretion of norepinephrine.

ANSWER: F

26. (T/F): Secretion of insulin and glucagon increase with increasing levels of physical exercise.

27. (T/F): Increased secretion of cortisol during exercise results in decreased protein catabolism.

ANSWER: F

28. (T/F): Decreased secretion of cortisol during exercise results in increased protein catabolism.

ANSWER: F

29. (T/F): Increased secretion of cortisol during exercise results in increased protein catabolism.

ANSWER: T

30. (T/F): Excess use of synthetic anabolic steroids may result in gynecomastia in male athletes.

ANSWER: T

31. (T/F): Methandrostenolone is rapidly metabolized by the liver.

ANSWER: F

32. (T/F): Methandrostenolone binds to plasma proteins much better than does testosterone.

ANSWER: F

33. (T/F): Nandrolone decanoate requires hydrolysis by the liver in order to bind to androgen receptors of target cells.

ANSWER: T

34. (T/F): Red blood cell packing is also known as blood doping.

ANSWER: T

35. (T/F): Polycythemia results from athletic training at high altitudes.

ANSWER: T

36. (T/F): Polycythemia is a naturally occurring form of red blood cell packing.

ANSWER: T

37. (T/F): One of the complications of growth hormone abuse by athletes is a condition called acromegaly.

ANSWER: T

38. (T/F): A pain-reducing effect is also termed an analgesic effect.

ANSWER: T

39. (T/F): Bicarbonate loading is an attempt to reduce lactic acid accumulation during strenuous physical exercise.

ANSWER: T

40. (T/F): Endogenous opiods have a pain-enhancing effect when released during physical exercise.

ANSWER: F

41. (T/F): A 'runner's high' is thought to be due to the release of endogenous opiods during strenuous physical activity.

ANSWER: T

42. (T/F): Electrolyte imbalance is not one of the contraindications of excessive bicarbonate loading.

ANSWER: F

43. (T/F): Diarrhea may result from excessive bicarbonate loading.

ANSWER: T

44. (T/F): Naturally occurring androgens are rapidly metabolized by the liver.

ANSWER: T

45. (T/F): Gynecomastia may result from excess use of synthetic anabolic steroids.

ANSWER: T

46. (T/F): Blood glucose and insulin levels increase as exercise increases in intensity and duration.

ANSWER: F

47. (T/F): Blood glucose and insulin levels decrease as exercise increases in intensity and duration.

ANSWER: T

48. (T/F): Cardiovascular function is not influenced by catecholamines during physical exercise.

ANSWER: F

49. (T/F): The heart rate of a well trained athlete is higher than that of the untrained person at rest.

ANSWER: F

50. (T/F): The stroke volume of the trained athlete is lower than that of the untrained person during exercise.

ANSWER: F

51. (T/F): Blood flow to the skin is enhanced during the very early stages of physical exercise.

ANSWER: F

52. (T/F): Blood flow to the skin is diminished during the very early stages of physical exercise.

ANSWER: T

53. (T/F): An increase of 5 degrees centigrade over the resting body temperature can lead to symptoms of heat stroke.

ANSWER: T

54. (T/F): There is a near linear relationship between the increase in body temperature and the percent of maximum oxygen consumption experienced during physical exercise.

55. (T/F): The slow component of the oxygen debt is also known as the alactic acid oxygen debt.

ANSWER: F

56. (T/F): At rest the MRV is on the order of approximately 2 to 3 liters.

ANSWER: F

57. (T/F): Tidal volume may increase to 2.0 liters or greater during physical exercise.

58. (T/F): Use of anabolic steroids also has an important psychological component for certain athletes.

ANSWER: T

59. (T/F): Long-term use of synthetic anabolic steroids has been linked to an increased incidence of atherosclerosis and coronary heart disease.

ANSWER: T

60. (T/F): Pulmonary blood flow increases with increasing physical activity.

ANSWER: T

61. (T/F): Increased pulmonary blood flow increases the surface area available for gaseous exchange of oxygen and carbon dioxide.

ANSWER: T

62. (T/F): Storgae of energy within muscle tissue occurs in the form of ATP.

ANSWER: F

63. (T/F): Storage of energy within muscle tissue occurs in the form of creatine phosphate.

ANSWER: T

64. (T/F): During the early stages of exercise, CP and the already present ATP are utilized for energy production.

ANSWER: T

65. (T/F): The amount of ATP present in muscle exceeds that of CP by two- or three-fold.

ANSWER: F

66. (T/F): The amount of CP present in muscle exceeds that of ATP by two- or three-fold.

ANSWER: T

67. (T/F): Maximal oxygen consumption is influenced by one's age and sex.

ANSWER: T

68. (T/F): Increased cardiac output is thought to be chiefly responsible for the increased maximal oxygen consumption observed with physical training.

ANSWER: T

69. (T/F): During strenuous exercise, the level of aerobic energy metabolism is markedly elevated over that of the anaerobic pathway.

ANSWER: F

70. (T/F): An increase in an athlete's MRV may be due initially to a psychological component.

ANSWER: T